Voice over LTE

Table of Contents

Preface

This book discusses the mechanisms used in the 4G EPS (Evolved Packet System) mobile network for telephone service support, and in the IMS network (IP (Internet Protocol) Multimedia Sub-system) to provide a telephone service.

The 4G network does not provide a telephone service because it does not process telephone signaling. It operates in PS (Packet Service) mode, and only transports IP packets. Therefore, it only transfers IP packets containing voice data or telephone signaling.

The IP packet containing voice data has the following structure:

– AMR (Adaptative Multi-Rate) codec;

– RTP (Real-Time Transport Protocol) header;

– UDP (User Datagram Protocol) header;

– IP header.

The IP packet containing telephone signaling has the following structure:

– SIP (Session Information Protocol) message;

– UDP header;

– IP header.

Chapter 1 introduces the different entities of the 4G network. It describes the 4G signaling protocol exchanged between the different entities, enabling a mobile to attach, update its location, establish sessions for the transport of IP packets and change cells (known as handover). For the purposes of the transport of IP packets, the 4G network has supports in place that are known as bearers. A bearer is similar to a virtual circuit.

Each bearer has a QCI (QoS Class Identifier) associated with it. Thus, for each mobile, two bearers are created: one for the transport of the telephone signaling (QCI = 5) and the other for the transport of the voice data (QCI = 1).

Chapter 2 presents the LTE (Long Term Evolution) radioelectric interface between the mobile and the 4G networks. The radioelectric interface serves to transport the mobile traffic (IP packets containing voice data or telephone signaling) and the 4G signaling exchanged with the 4G network.

The procedures specific to the radioelectric interface relate to connection of the mobile to the 4G network, scheduling of the IP packets and re-transmission in the case of error.

To begin with, the establishment of a telephone communication will not be done over a 4G network, because of the difficulty in handover from PS mode to CS (Circuit Service) mode, when the mobile is transferred from a 4G cell to a 2G or 3G cell.

Chapter 3 discusses the mechanism of CSFB (CS FallBack), which is an interim solution. It enables a mobile connected to the 4G network to receive an alert sent by a 2G/3G network (this is known as paging). This page is sent when a call comes in on the 2G/3G network. On receiving the page, the mobile is transferred to the 2G/3G network, over which the telephone communication can then be established.

Similarly, a mobile connected to the 4G network and wishing to make an outgoing call must first be transferred to the 2G/3G network.

Chapter 4 presents the SIP protocol, upon which the telephone signaling transferred by the 4G network is based. SIP defines two fundamental procedures: registration of the mobile and establishment of the session (i.e. the telephone communication).

Chapter 5 introduces the IMS network which provides a telephone service, using the 4G network for the transport of the voice data and telephone signaling. The telephone signaling again uses SIP, enriching it.

The IMS network defines the routing of the telephone signaling, access to databases containing the profile and secret data of the subscriber, and the specific processing of voice data to provide particular services, such as conference calling, for instance.

Telephone communication can be established between two 4G mobiles. The telephone signaling is processed by the IMS entities of the home operator of each mobile. The voice data is directly transferred between the 4G networks (see Figure 1).

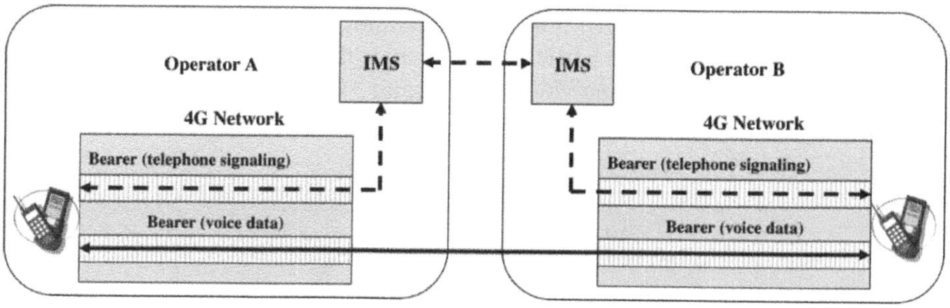

Figure 1. *Telephone communication between two 4G mobiles*

Telephone communication can also be established between a mobile and a terminal connected to the fixed network PSTN (Public Switched Telephone Network) or the mobile network PLMN (Public Land Mobile Network). The IMS network provides the entities which perform conversion of the protocols and interconnection with these networks (Figure 2).

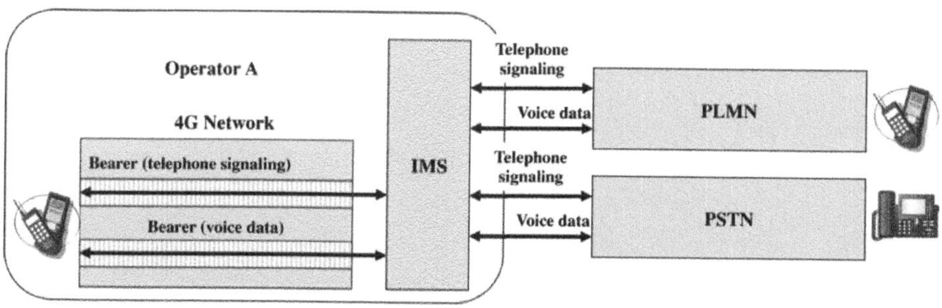

Figure 2. *Telephone communication between a 4G mobile and a terminal connected to the PLMN or PSTN network*

Chapter 6 presents the telephone services offered by a particular entity within the IMS network – the TAS (Telephone Application Server). These services relate to communication forwarding, identification presentation or restriction, message waiting indication, communication hold, conference, communication transfer, call waiting, malicious communication identification, completion of communication, call rejection and announcements.

The telephone communication established over the 4G network in PS mode needs to be maintained when the mobile is transferred to the 2G/3G network in CS mode.

Chapter 7 finally discusses the mechanism of SRVCC (Single Radio Voice Call Continuity), which takes care of this call maintaining in the case of a PS–CS inter-system handover. SRVCC is a particular function of the IMS network. It anchors the flows of telephone signaling and voice data (Figure 3).

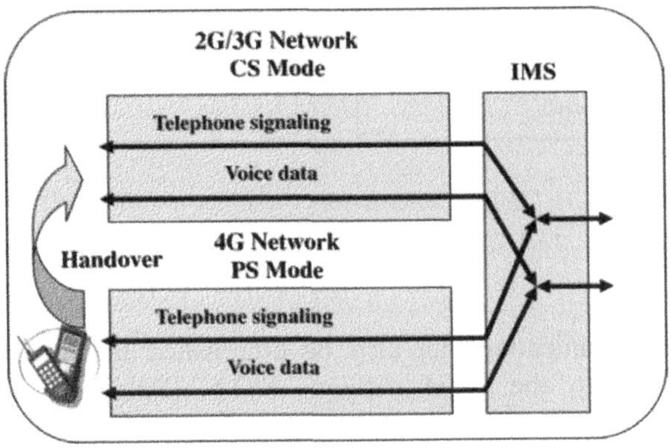

Figure 3. *The SRVCC mechanism*

Acronyms

A

AAA	Authorization-Authentication-Answer
AAL2	ATM Adaptation Layer 2
AAR	Authorization-Authentication-Request
ACM	Address Complete Message
ACR	Anonymous Communication Rejection
AM	Acknowledged Mode
AMR	Adaptive Multi-Rate
ANM	Answer Message
APN	Access Point Name
ARQ	Automatic Repeat reQuest
AS	Application Server
ASA	Abort-Session-Answer
ASR	Abort-Session-Request
ATCF	Access Transfer Control Function
ATGW	Access Transfer Gateway
ATM	Asynchronous Transfer Mode

B

B2BUA	Back-to-Back User Agent
BCCH	Broadcast Control Channel
BCH	Broadcast Channel
BGCF	Breakout Gateway Control Function

BICC	Bearer Independent Call Control
BSS	Base Station Sub-system
BSSMAP	BSS Management Application Part

C

CCBS	Completion of Communications to Busy Subscriber
CCCH	Common Control Channel
CCNL	Completion of Communications on Not Logged-in
CCNR	Completion of Communications on No Reply
CD	Communication Deflection
CDF	Charging Data Function
CDIV	Communication Diversion
CDR	Charging Data Record
CFB	Communication Forwarding on Busy user
CFI	Control Format Indicator
CFNL	Communication Forwarding on Not Logged-in
CFNR	Communication Forwarding on No Reply
CFU	Communication Forwarding Unconditional
CGF	Charging Gateway Function
CM	Call Management
CQI	Channel Quality Indicator
C-RNTI	Cell Radio Network Temporary Identity
CS	Circuit Service
CSCF	Call Session Control Function
CSFB	Circuit Service FallBack
CTF	Charging Trigger Function
CW	Communication Waiting

D

DCCH	Dedicated Control Channel
DCI	Downlink Control Information
DFTS	Discrete Fourier Transform Spread
DL-SCH	Downlink Shared Channel
DNS	Domain Name System
DRB	Data Radio Bearer

DRS	Demodulation Reference Signal
DSCP	DiffServ Code Point
DTCH	Dedicated Traffic Channel
DTM	Dual Transfer Mode
DwPTS	Downlink Pilot Time Slot

E

E-CSCF	Emergency-CSCF
ECT	Explicit Communication Transfer
EMM	EPS Mobility Management
eNB	evolved Node B
EPC	Evolved Packet Core
EPS	Evolved Packet System
E-RAB	EPS Radio Access Bearer
ESM	EPS Session Management
ETWS	Earthquarke and Tsunami Warning System
E-UTRAN	Evolved Universal Terrestrial Radio Access Network

F

FDD	Frequency Division Duplex

G

GMSC	Gateway MSC
GP	Gap Period
GPRS	General Packet Radio Service
GSM	Global System for Mobile
GTP-C	GPRS Tunnel Protocol Control
GTP-U	GPRS Tunnel Protocol User
GUTI	Globally Unique Temporary Identity

H

HARQ	Hybrid ARQ
HDB3	High Density Binary 3

HI	HARQ Indicator
HSS	Home Subscriber Server
HTTP	Hypertext Transfer Protocol

I

IAM	Initial Address Message
ICB	Incoming Communication Barring
ICIC	Inter-Cell Interference Coordination
I-CSCF	Interrogating-CSCF
iFC	initial Filter Criteria
IFFT	Inverse Fast Fourier Transform
IMS	IP Multimedia Sub-system
IMS-GWF	IMS Gateway Function
IMSI	International Mobile Subscriber Identity
IP	Internet Protocol
IPSec	IP Security
ISIM	IMS Services Identity Module
ISUP	ISDN User Part

L

LAI	Location Area Identifier
LIA	Location-Info-Answer
LIR	Location-Info-Request
LTE	Long Term Evolution

M

M3UA	MTP 3 User Adaptation
MAA	Multimedia-Auth-Answer
MAC	Media Access Control
MAR	Multimedia-Auth-Request
MCC	Mobile Country Code
MCCH	Multicast Control Channel
MCH	Multicast Channel

MCID	Malicious Communication Identification
MeGaCo	Media Gateway Controller
MGCF	Media Gateway Control Function
MGW	Multimedia Gateway
MIB	Master Information Block
MIMO	Multiple Input Multiple Output
MISO	Multiple Input Single Output
MM	Mobility Management
MME	Mobility Management Entity
MMEC	MME Code
MMEGI	MME Group Identity
MMEI	MME Identity
MNC	Mobile Network Code
MRF	Multimedia Resource Function
MRFC	MRF Controller
MFRP	MRF Processor
MSC	Mobile-services Switching Centre
MTCH	Multicast Traffic Channel
M-TMSI	MME - Temporary Mobile Subscriber Identity
MTP	Message Transfer Part
MU-MIMO	Multi User-MIMO
MWI	Message Waiting Indication

N

NAS	Non Access Stratum
NGN	Next Generation Network

O

OCB	Outgoing Communication Barring
OCS	Online Charging System
OFDM	Orthogonal Frequency-Division Multiplexing
OIP	Originating Identification Presentation
OIR	Originating Identification Restriction

P

PBCH	Physical Broadcast Channel
PCCH	Paging Control Channel
PCEF	Policy and Charging Enforcement Function
PCFICH	Physical Control Format Indicator Channel
PCH	Paging Channel
PCRF	Policy Charging and Rules Function
P-CSCF	Proxy-CSCF
PDCCH	Physical Downlink Control Channel
PDCP	Packet Data Convergence Protocol
PDN	Packet Data Network
PDSCH	Physical Downlink Shared Channel
PGW	PDN Gateway
PHICH	Physical HARQ Indicator Channel
PLMN	Public Land Mobile Network
PMCH	Physical Multicast Channel
PMI	Precoding Matrix Indicator
PPA	Push-Profile-Answer
PPR	Push-Profile-Request
PRACH	Physical Random Access Channel
PS	Packet Service
PSS	Primary Synchronization Signal
PSTN	Public Switched Telephone Network
PUCCH	Physical Uplink Control Channel
PUSCH	Physical Uplink Shared Channel

Q

QAM	Quadrature Amplitude Modulation
QCI	QoS Class Identifier
QoS	Quality of Service
QPSK	Quadrature Phase-Shift Keying

R

RACH	Random Access Channel

RANAP	Radio Access Network Application Part)
RAR	Random Access Response
RB	Resource Block
RE	Resource Element
REG	Resource Element Group
REL	Release
RI	Rank Indicator
RLC	Radio Link Control
RLC	Release Complete
ROHC	Robust Header Compression
RRC	Radio Resource Control
RS	Reference Signal
RSRP	Reference Signal Received Power
RSRQ	Reference Signal Received Quality
RTA	Registration-Termination-Answer
RTP	Real-Time Transport Protocol
RTR	Registration-Termination-Request

S

SAA	Server-Assignment-Answer
SAE	System Architecture Evolution
SAR	Server-Assignment-Request
SCC AS	Service Centralization and Continuity Application Server
SCCP	Signaling Connection Control Part
S-CSCF	Serving-CSCF
SCTP	Stream Control Transmission Protocol
SDH	Synchronous Digital Hierarchy
SDP	Session Description Protocol
SGSN	Service GPRS Support Node
SFBC	Space Frequency Block Code
SFN	System Frame Number
SGW	Serving Gateway
SGW	Signaling Gateway
SIB	System Information Block
SIGTRAN	Signaling Transport over IP
SIMO	Single Input Multiple Output

SIP	Session Information Protocol
SISO	Single Input Single Output
SLF	Subscription Locator Functional
SRB	Signaling Radio Bearer
SRS	Sounding Reference Signal
SRVCC	Single Radio Voice Call Continuity
SS7	Signaling System 7
SSS	Secondary Synchronization Signal
STA	Session-Termination-Answer
S-TMSI	Shortened-TMSI
STN-SR	Session Transfer Number for SRVCC
STR	Session-Termination-Request
SU-MIMO	Single User MIMO

T

TAC	Tracking Area Code
TAI	Tracking Area Identity
TAS	Telephony Application Server
TCP	Transmission Control Protocol
TDD	Time Division Duplex
TDM	Time Division Multiplexing
TEID	Tunnel Endpoint Identifier
TIP	Terminating Identification Presentation
TIR	Terminating Identification Restriction
TLS	Transport Layer Security
TM	Transparent Mode
TMSI	Temporary Mobile Subscriber Identity
TTI	Transmission Time Interval

U

UA	User Agent
UAA	User-Authorization-Answer
UAC	User Agent Client
UAR	User-Authorization-Request
UAS	User Agent Server

UCI	Uplink Control Information
UDP	User Datagram Protocol
UE	User Equipment
UL-SCH	Uplink Shared Channel
UM	Unacknowledged Mode
UMTS	Universal Mobile Telecommunications System
UpPTS	Uplink Pilot Time Slot
URI	Uniform Resource Identifier
UTRAN	Universal Terrestrial Radio Access Network

X

XML	eXtensible Markup Language

Chapter 1

The EPS Network

1.1. Architecture

The 4th-generation mobile network EPS (Evolved Packet System) comprises a core network EPC (Evolved Packet Core) and an access network E-UTRAN (Evolved Universal Terrestrial Radio Access Network) (Figure 1.1).

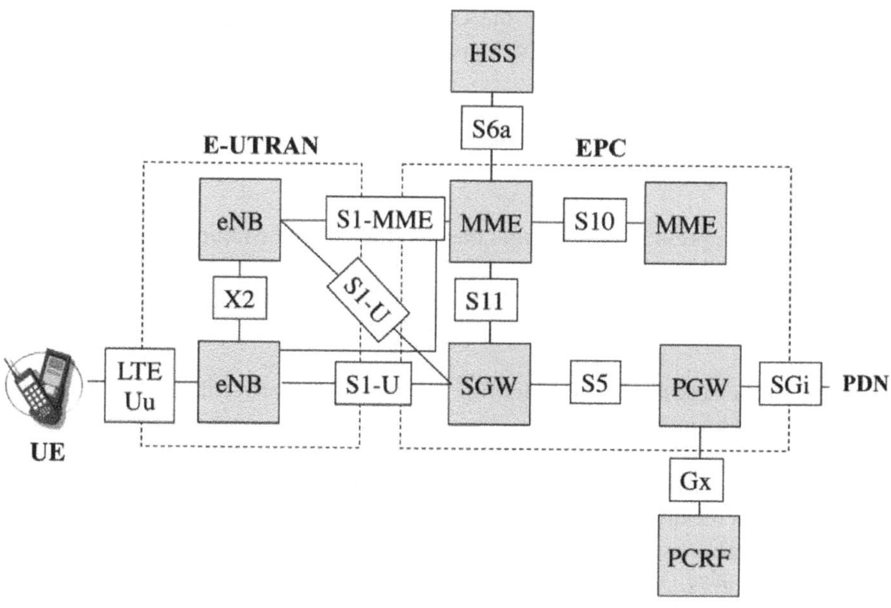

Figure 1.1. *Architecture of the EPS network*

The access network E-UTRAN takes care of connection of mobiles. The EPC core network interconnects the access network and provides the interface for the PDN (Packet Data Network). It ensures the attachment of mobiles to the network and the establishment of the bearers.

The term SAE (System Architecture Evolution) is used for the study of the evolution of the core network EPC.

The term LTE (Long Term Evolution) is attributed to the study of the evolution of the radioelectric interface Uu between the EPS and the mobile UE (User Equipment).

1.1.1. *Access network*

The access network E-UTRAN includes only one type of entity, the radioelectric station eNB (evolved Node B) to which the UE connects (Figure 1.1).

The eNB is responsible for managing radioelectric resources, the allocation of bearers to the mobile and the mobility of the UE.

The eNB transfers the traffic data from the mobile (or respectively the SGW (Serving Gateway) of the EPC) to the SGW (or respectively the mobile). When the eNB receives data from the UE or from the SGW, it examines the QCI (QoS Class Identifier) to implement the packet scheduling mechanism.

For outgoing data destined for the SGW, the eNB performs DSCP (DiffServ Code Point) marking of the IP (Internet Protocol) packets in relation with the QCI assigned to each packet.

The eNB compresses and encrypts the data traffic on the radioelectric interface.

The eNB encrypts and controls the integrity of the signaling data exchanged with the mobile.

The eNB selects the MME (Mobility Management Entity) in the EPC to which the mobile will be attached.

The eNB processes the paging request sent by the MME for broadcast into the cell. The cell is the area of the eNB's radioelectric coverage.

The eNB also broadcasts the data relating to the characteristics of the radioelectric interface into the cell, which the mobile uses to connect.

The eNB uses the measurements carried out by the mobile to make a decision on whether, or not, to trigger a handover and for scheduling the data packets exchanged with the mobile.

The eNB has the following interfaces (Figure 1.1):

– LTE-Uu with the mobile UE. This interface is used to connect the mobile to the eNB. It carries traffic from the mobile and the signaling exchanged between the mobile and the eNB. This signaling supports the signaling exchanged between the mobile and the MME of the EPC.

– X2 with the other eNBs. This interface is used for intra-E-UTRAN handover and for exchanging cell load information. It carries mobile traffic and the signaling exchanged between two eNBs.

– S1-MME with the MME of the EPC. This interface is used for activating the radioelectric bearer, for paging and for mobility management. It carries the signaling exchanged between the MME and the eNB. This signaling carries the signaling exchanged between the mobile and the MME.

– S1-U with the SGW of the EPC. This interface only carries the mobile traffic.

1.1.2. *Core network*

The EPC is made up of the MMEs only performing processing of the signaling from the EPS, and the SGW and PGW (PDN Gateway), which handle the transfer of the traffic data (Figure 1.1).

1.1.2.1. *The MME*

The MME is the control tower of the EPS. It grants mobiles access to the EPS and controls the establishment of bearers for transmission of the traffic data.

The MMEs belong to a pool. The load balancing of the MMEs is managed by the eNBs which need access to each of the MMEs in the same pool.

Each MME is identified by the parameters MMEGI (MME Group Identity) and MMEC (MME Code), which together make up the MMEI (MME Identity).

The MME is responsible for attachment and detachment of the mobile.

When an attachment is established, the MME retrieves the profile and authentication data for the mobile stored on the HSS (Home Subscriber Server), and authenticates the mobile. The HSS is a database which can be shared between the different generations of mobile networks and the IMS (IP Multimedia Sub-system) network.

When an attachment is established, the MME registers the TAI (Tracking Area Identity) of the mobile and assigns the UE a GUTI (Globally Unique Temporary Identity).

The TAI comprises the following fields:

– MCC (Mobile Country Code);

– MNC (Mobile Network Code);

– TAC (Tracking Area Code).

The GUTI comprises the following fields:

– MCC;

– MNC;

– the MMEI of the MME to which the mobile is attached;

– the M-TMSI (MME – Temporary Mobile Subscriber Identity) assigned to the mobile by the MME.

The S-TMSI (Shortened-TMSI) is composed by combining the M-TMSI and the MMEC of the MME. The MME uses the S-TMSI for paging.

The MME manages a list of the TAIs allocated to mobiles, within which the mobile, in idle mode, can move without contacting the MME to update its tracking area.

When a mobile attaches, the MME selects the SGW and PGW to form the default bearer. The default bearer is formed when an attachment is established by the mobile (e.g. for the bearer of telephone signaling).

The MME also processes the mobile request for the establishment, modification and release of a dedicated bearer for the transmission of traffic data. The dedicated bearer is constructed on request from the mobile for a particular service (for instance, for voice data transmission).

For the construction of the bearer, the selection of the PGW is carried out on the basis of the APN (Access Point Name) communicated by the mobile or by the HSS in the subscriber's profile.

The MME also selects the target MME in the case of the handover of the mobile from one pool to another.

The MME provides the data required for legal interceptions, such as the state of the mobile (idle or connected), its TAI and the characteristics of the traffic.

The MME has the following interfaces (Figure 1.1):

– S1-MME with the eNB;

– S6a with the HSS. This interface carries the signaling in order to access the mobile data (authentication, service profile);

– S10 with the MME. This interface carries the signaling exchanged when the UE moves and necessitates a switch of MME;

– S11 with the SGW. This interface carries the signaling allowing the establishment of the bearer between the eNBs and the SGW.

1.1.2.2. *The SGW*

SGWs are also organized in pools. To balance the load of the SGWs, all the eNBs in an area must have access to all of the SGWs in the same pool.

The SGW transfers the incoming data from the PGW to the eNB and the outgoing data from the eNB to the PGW. When the SGW receives data from the eNBs or PGWs, it examines the QCI in order to implement the packet scheduling mechanism.

For the incoming and outgoing data, the SGW performs DSCP marking of the IP packets on the basis of their assigned QCI.

The SGW orders the traffic data when an inter E-UTRAN handover occurs.

The SGW is the anchor point for intra-system handover (within the 4G network), provided the mobile does not change pool. Otherwise, it is the PGW that performs this function.

The SGW is the anchor point for inter-system handover in PS mode, necessitating the transfer of traffic from the mobile to a 2G or 3G mobile network.

The SGW initiates notification to the MME for incoming data when the mobile is in idle mode. A mobile in idle mode remains attached to the MME. However, it is no longer connected to the eNB, so the bearers are released.

The SGW has the following interfaces (Figure 1.1):

– S11 with the MME;

– S5 with the PGW. This interface is used to establish a bearer between these two entities. It carries the signaling exchanged with the PGW and the mobile traffic;

– S1-U with the eNB.

1.1.2.3. *The PGW*

The PGW is the gateway router connecting the EPS to the PDN (i.e. the Internet network).

When the PGW receives data from the SGW or from the PDN, it uses the QCI to implement the packet-scheduling mechanism.

The PGW performs DSCP marking of the IP packets on the basis of their assigned QCI.

When an attachment is established, the PGW assigns a IPv4 or IPv6 address to the mobile.

The PGW is the anchor point for inter-SGW switch.

The PGW contains the PCEF (Policy and Charging Enforcement Function), which applies the rules relating to mobile traffic, packet filtering, charging and the QoS to be applied to the bearer being constructed. The PCRF (Policy Charging and Rules Function), outside of the EPS, tells the PCEF of the PGW which rules need to be applied.

The PGW generates data enabling charging entities to edit the charging records which are dealt with by the billing system.

The PGW diverts the traffic from the mobile in the case of legal interceptions.

The PGW has the following interfaces (Figure 1.1):

– S5 with the PGW;

– Gx with the PCRF. This interface carries the signaling enabling the PGW to receive the rules to be applied to the mobile traffic;

– SGi with the PDN. This interface carries the mobile traffic (IP packet).

1.1.3. *Protocol architecture*

The LTE-Uu interface is the reference point between the mobile and the eNB for the signaling and traffic. RRC (Radio Resource Control) signaling is exchanged between the mobile and the eNB. The RRC protocol also deals with the transport of the NAS (Non Access Stratum) protocol exchanged between the mobile and the MME (Figure 1.2).

The S1-MME interface is the reference point between the MME and eNB for signaling using the S1-AP (Application Part) protocol. S1-AP also deals with the transport of the NAS protocol exchanged between the mobile and the MME (Figure 1.2).

The S11 interface is the reference point between the MME and SGW for signaling using the GTP-C [GPRS (General Packet Radio Service) Tunnel Protocol Control] protocol (Figure 1.2).

The S1-U interface is the reference point between the eNB and SGW for tunneling of the traffic (the IP packet) via the GTP-U (GPRS Tunnel Protocol User) protocol (Figure 1.3).

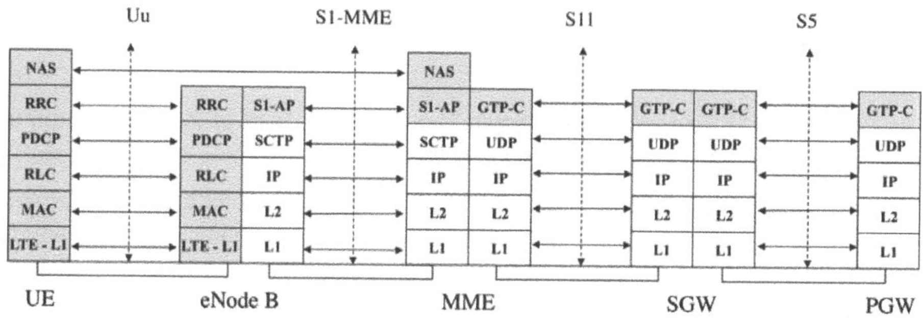

The shaded blocks are described in this book.
L2 (Layer 2): data link layer
L1 (Layer 1): physical layer

Figure 1.2. *Protocol architecture – the control plane*

The S5 interface is the reference point between the SGW and PGW for tunneling the traffic (the IP packet) via the GTP-U protocol (Figure 1.3) and for signaling via the GTP-C protocol (Figure 1.2).

The S10 interface is the reference point between the MMEs for signaling using GTP-C (Figure 1.1).

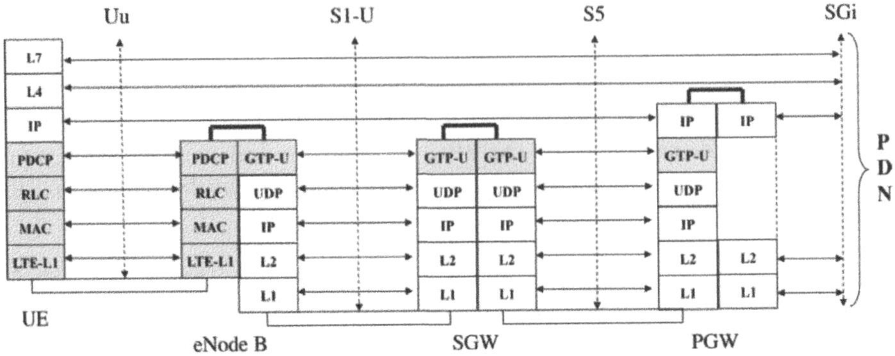

The shaded blocks are described in this book.
L7 (Layer 7): application layer
L4 (Layer 4): transport layer

Figure 1.3. *Protocol architecture – the traffic plane*

The SGi interface is the reference point between the PDW and the PDN (Figure 1.3).

The EPS transports mobile traffic (an IP packet) transparently to the PGW, which routes that packet. To perform this function, the network entities create bearers:

– the LTE-Uu Bearer (Radio Bearer). The eNB takes care of the construction of this bearer using RRC signaling exchanged with the mobile;

– the S1-U Bearer is identified by the TEID (Tunnel Endpoint Identifier), carried by the GTP-U protocol. The MME constructs this bearer using S1-AP signaling exchanged with the eNB and GTP-C exchanged with the SGW;

– the S5 Bearer is identified by the TEID carried by GTP-U. The SGW constructs this bearer using GTP-C signaling exchanged with the PGW. The identity of the PGW is communicated to the SGW by the MME.

To transfer the traffic data, the entities in the EPS network use a lookup table between the bearer IDs. In fact, the EPS constructs a virtual circuit (EPS Bearer) between the mobile and the PGW. Each EPS Bearer is attributed a QCI value.

The combination of the Radio Bearer and S1-U Bearer constitutes the E-RAB (EPS Radio Access Bearer).

Transport of the mobile traffic and the signaling between the entities in the EPS is taken care of by an IP network for which the entities in the EPS represent hosts.

At the LTE-Uu radioelectric interface, the signaling and the traffic from the mobile are borne by a data link layer and a physical layer.[1]

The data link layer is structured into three sub-layers:

– PDCP (Packet Data Convergence Protocol). This protocol takes care of the compression of the traffic data, the encryption of the traffic and signaling data, the control of the integrity of the signaling data and the scheduling of the traffic data when an intra-system handover occurs;

– RLC (Radio Link Control). This implements the retransmission mechanism in the case of error ARQ (Automatic Repeat reQuest), and concatenates or segments the PDCP data;

– MAC (Media Access Control). This protocol performs the multiplexing of the RLC data, implements the HARQ (Hybrid ARQ) mechanism in the case of error and schedules the downlink and uplink data.

The physical layer determines the characteristics of transmission over the LTE-Uu radioelectric interface (Table 1.1).

Frequency bandwidth	1.4 MHz, 3, 5, 10, 15, 20 MHz
Principle of duplex between both transmission directions	TDD or FDD
Modulation of sub-carriers	Downstream: QPSK, 16-QAM, 64-QAM Upstream: QPSK, 16-QAM, 64-QAM (for a particular category of mobile)
Principle of multiplexing of sub-carriers	Downstream: OFDM Upstream: DFTS-OFDM
Antenna system	MISO, MIMO
Maximum throughput	Downstream: 300 Mbit/s (Note 1) Upstream: 75 Mbit/s (Note 2)
Multiple access	Random attribution of resource blocks: 0.5 ms time and 180 kHz frequency

TDD: Time Division Duplex
FDD: Frequency Division Duplex
QPSK: Quadrature Phase-Shift Keying
QAM: Quadrature Amplitude Modulation
OFDM: Orthogonal Frequency-Division Multiplexing
DFTS: Discrete Fourier Transform Spread
MISO: Multiple Input Single Output
MIMO: Multiple Input Multiple Output
Note 1: the maximum throughput is obtained for a bandwidth of 20 MHz, 64-QAM modulation and a 4×4 MIMO antenna system.
Note 2: the maximum throughput is obtained for a bandwidth of 20 MHz and 64-QAM modulation.

Table 1.1. *Characteristics of the physical layer of the LTE-Uu interface*

1 A detailed description of the data link layer and the physical layer is given in Chapter 2, on the LTE interface.

The X2 interface is the reference point between two eNBs for signaling using the X2-AP protocol (Figure 1.4) and for tunneling of the mobile traffic (the IP packet) via the GTP-U protocol, during a handover of the mobile (Figure 1.5).

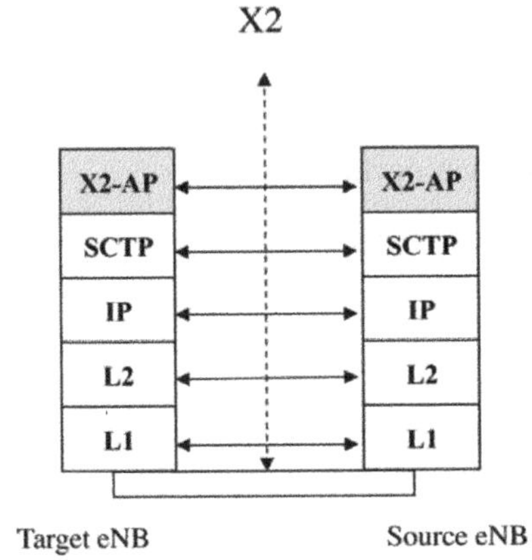

Figure 1.4. *Protocol architecture over the X2 interface – the control plane*

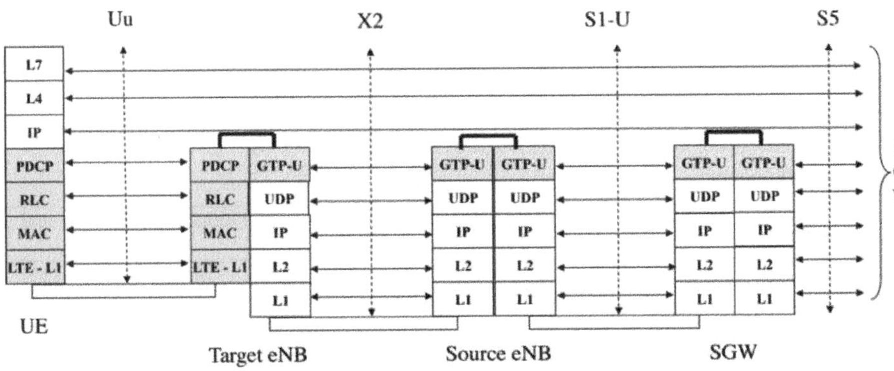

Figure 1.5. *Protocol architecture – the traffic plane during handover based on the X2 interface*

The tunnel established between the two eNBs is unidirectional (source eNB to target eNB). It is used to transfer the data traffic received from the SGW to the target eNB. It is established temporarily, while the handover of the mobile takes place.

The S6a interface is the point of interface between the MME and HSS for signaling via the protocol DIAMETER, facilitating access to the mobile data (authentication, service profile) (Figure 1.1).

The Gx interface is the reference point between the PCRF and PGW for signaling via DIAMETER with regard to transfer of filtering-, QoS- and charging-rules (Figure 1.1).

1.2. Signaling protocols

1.2.1. *NAS protocol*

The NAS protocol is the signaling exchanged between the mobile and the MME. It is transported by the RRC protocol over the radioelectric interface LTE-Uu and by the S1-AP protocol over the S1-MME interface (Figure 1.2). It comprises the following two protocols:

– EMM (EPS Mobility Management). This takes care of controlling mobility and security;

– ESM (EPS Session Management). This controls the bearer establishment.

From the point of view of the MME, the mobile can be in one of two operational states: EMM-REGISTERED or EMM-DEREGISTERED.

In the EMM-DEREGISTERED state, the mobile's location is not known to the MME and, therefore, it cannot be contacted. The switch to the registered state takes place when the mobile attaches, which comprises the following four procedures:

– mutual authentication of the mobile and the MME;

– registration of the mobile's location with the MME;

– assignment of the GUTI to the mobile;

– establishment of the default bearer.

The switch to the deregistered state takes place when the mobile detaches or when the attachment of the mobile, the update of its location or the service request are rejected by the MME.

1.2.1.1. *EMM messages*

1.2.1.1.1. Attachment and detachment

The procedure of attachment is initiated by the mobile in the deregistered state, by sending the message EMM ATTACH REQUEST to the MME. This message contains the mobile GUTI or IMSI and its TAC. The mobile attaches the message ESM PDN CONNECTIVITY REQUEST to establish the default bearer.

Upon receiving this message, the MME begins the authentication and security procedures. If they are successfully completed, the MME responds with the message EMM ATTACH ACCEPT, containing a new GUTI, and the message ESM ACTIVATE DEFAULT EPS BEARER CONTEXT REQUEST, to establish the default bearer.

The mobile responds with the message EMM ATTACH COMPLETE containing the message ESM ACTIVATE DEFAULT EPS BEARER CONTEXT ACCEPT, to acknowledge the previous message.

If the procedures are not successful, the MME responds with the message EMM ATTACH REJECT, containing the message ESM PDN CONNECTIVITY REJECT, which causes the mobile to detach.

Detachment may be initiated by the mobile or the MME by sending the message EMM DETACH REQUEST. The response EMM DETACH ACCEPT concludes the detachment procedure. The response is not transmitted by the MME if the detach request sent by the mobile indicates that it has been turned off. The detachment procedure implicitly causes the release of the active bearers.

1.2.1.1.2. Authentication

The procedure of mutual authentication is initiated by the MME by sending the message EMM AUTHENTICATION REQUEST, containing a random number RAND and the authentication code AUTN (Authentication Network).

The mobile uses the RAND received to locally compute its own authentication code RES (Result), and compares its AUTN to the one received from the MME.

If the MME is authenticated, the mobile responds with the message EMM AUTHENTICATION RESPONSE, containing the authentication code RES. Otherwise, it displays the message EMM AUTHENTICATION FAILURE.

The MME compares the RES value received from the mobile with that communicated by the HSS. If the two codes are the same, the mobile is authenticated and the MME triggers security mode for NAS signaling. Otherwise, it responds with the message EMM AUTHENTICATION REJECT.

1.2.1.1.3. Security mode

When mutual authentication has been successful, the MME begins putting the NAS signaling in security mode by sending the message EMM SECURITY MODE COMMAND. The integrity of this message is protected.

If the check on the integrity of the EMM SECURITY MODE COMMAND is positive, the mobile responds with the message EMM SECURITY MODE COMPLETE. All subsequent NAS messages are encrypted and their integrity is checked.

If the check on the integrity of the EMM SECURITY MODE COMMAND is negative, the mobile responds with the message EMM SECURITY MODE REJECT.

1.2.1.1.4. Tracking area update

The procedure of tracking area update is periodically initiated so that the mobile can maintain its tracking area, or when the mobile has changed TAC. The mobile, in the registered state, sends the message EMM TRACKING AREA UPDATE REQUEST to the MME.

The MME responds either with the message EMM TRACKING AREA UPDATE ACCEPT if it accepts the update, or else with the message EMM TRACKING AREA UPDATE REJECT, indicating the cause of the rejection.

If the message EMM TRACKING AREA UPDATE ACCEPT attributes a new GUTI to the mobile, the mobile confirms receipt of this by sending the message EMM TRACKING AREA UPDATE COMPLETE.

1.2.1.1.5. Service request

The service request is initiated by the mobile, by sending the EMM SERVICE REQUEST when signaling or traffic data is waiting. The mobile is notified of awaiting data at the level of the network by way of the paging procedure. The service request is sent when the mobile is in idle mode, to re-establish the bearers on the LTE-Uu and S1-U interfaces.

The MME may reject the request, in which case it responds with the message EMM SERVICE REJECT. This response causes the passage of the mobile to the deregistered state.

1.2.1.2. *ESM messages*

The mobile sends the request to establish the default bearer when the mobile attaches to the MME. The dedicated bearer corresponds to a specific request by the mobile. The dedicated bearer is associated with a particular quality of service, corresponding to a particular QCI, which is different to that of the default bearer. The establishment request can be transmitted by the network or the mobile.

Table 1.2 recaps the ESM messages exchanged for the establishment, modification and release of the bearer.

Source	Message	Destination
	Establishment of default bearer, initiated by the network	
MME	ACTIVATE DEFAULT EPS BEARER CONTEXT REQUEST	UE
UE	ACTIVATE DEFAULT EPS BEARER CONTEXT ACCEPT or ACTIVATE DEFAULT EPS BEARER CONTEXT REJECT	MME
	Establishment of dedicated bearer, initiated by the network	
MME	ACTIVATE DEDICATED EPS BEARER CONTEXT REQUEST	UE
UE	ACTIVATE DEDICATED EPS BEARER CONTEXT ACCEPT or ACTIVATE DEDICATED EPS BEARER CONTEXT REJECT	MME
	Modification of dedicated bearer, initiated by the network	
MME	MODIFY EPS BEARER CONTEXT REQUEST	UE
UE	MODIFY EPS BEARER CONTEXT ACCEPT or MODIFY EPS BEARER CONTEXT REJECT	MME

Table 1.2. *Messages in the ESM protocol*

Release of dedicated bearer, initiated by the network		
MME	DEACTIVATE EPS BEARER CONTEXT REQUEST	UE
UE	DEACTIVATE EPS BEARER CONTEXT ACCEPT	MME
Release of default bearer, initiated by the mobile		
UE	PDN CONNECTIVITY REQUEST	MME
MME	ACTIVATE DEFAULT EPS BEARER CONTEXT REQUEST or PDN CONNECTIVITY REJECT	UE
Establishment or modification of dedicated bearer, initiated by the mobile		
UE	BEARER RESOURCE ALLOCATION REQUEST	MME
MME	ACTIVATE DEDICATED EPS BEARER CONTEXT REQUEST or ACTIVATE DEDICATED EPS BEARER CONTEXT REJECT or MODIFY EPS BEARER CONTEXT REQUEST	UE
Modification of dedicated bearer, initiated by the mobile		
UE	BEARER RESOURCE MODIFICATION REQUEST	MME
MME	ACTIVATE DEDICATED EPS BEARER CONTEXT REQUEST or MODIFY EPS BEARER CONTEXT REQUEST or DEACTIVATE EPS BEARER CONTEXT REQUEST or BEARER RESOURCE MODIFICATION REJECT	UE
Release of default bearer, initiated by the mobile		
UE	PDN DISCONNECT REQUEST	MME
MME	DEACTIVATE EPS BEARER CONTEXT REQUEST or PDN DISCONNECT REJECT	UE

Table 1.2. *(Continued) Messages in the ESM protocol*

1.2.2. *RRC protocol*

The RRC protocol is the signaling exchanged between the mobile and the eNB over the LTE-Uu radioelectric interface (Figure 1.2). It performs the following functions:

– broadcast of system information relative to the characteristics of the radioelectric interface;

– control of the RRC connection. This procedure includes the procedures of paging, establishment, modification and release of the SRB (Signaling Radio Bearer) and the DRB (Data Radio Bearer). It also includes the activation of security mode, the procedure for which consists of putting mechanisms in place to encrypt the traffic and signaling flows, and to control the integrity of the signaling flows;

– control of handover. This procedure executes the changing of cell between two eNBs (intra-system handover) or between an eNB and an entity from a 2nd- or 3rd-generation mobile network (inter-system handover);

– measurement reporting. The eNB can trigger measurements carried out by the mobile, either periodically or on demand, to prepare for handover;

– transport of the NAS messages exchanged between the mobile and the MME.

From the point of view of the eNB, the mobile may be in one of two operational states: idle mode (RRC_IDLE) or connected mode (RRC_CONNECTED).

In idle mode, the mobile is not known to the eNB. It remains in this state until the RRC connection setup procedure is completed. The setup procedure is triggered by the mobile when it wishes to transmit traffic or signaling data. In that state, the mobile used the SRB0 bearer.

In connected mode, the mobile can transmit and receive signaling and traffic data. The mobile is attributed an identifier which is unique to the cell – the C-RNTI (Cell Radio Network Temporary Identity). In this state, the mobile uses either the SRB1 bearer for RRC messages with possible associated NAS messages, or the SRB2 bearer for RRC messages transporting solely NAS messages.

Table 1.3 recaps the type of SRB, the mode of RLC protocol and the channels used by the different RRC messages over the radioelectric interface.[2]

2 A detailed description of the data link layer and the physical layer is given in Chapter 2, on the LTE interface.

SRB	RLC mode	Logic channel	Transport channel	Physical channel
		MasterInformationBlock		
None	TM	BCCH	BCH	PBCH
		SystemInformationBlock		
None	TM	BCCH	DL-SCH	PDSCH
		Paging		
None	TM	PCCH	PCH	PDSCH
		RRCConnectionSetup *RRCConnectionReject* *RRCConnectionReestablishment* *RRCConnectionReestablishmentReject*		
SRB0	TM	CCCH	DL-SCH	PDSCH
		RRCConnectionRequest *RRCConnectionReestablishmentRequest*		
SRB0	TM	CCCH	UL-SCH	PUSCH
		RRCConnectionReconfiguration *RRCConnectionRelease* *SecurityModeCommand*		
SRB1	AM	DCCH	DL-SCH	PDSCH
		DLInformationTransfer (1)		
SRB2	AM	DCCH	DL-SCH	PDSCH
		RRCConnectionSetupComplete *SecurityModeComplete* *SecurityModeFailure* *RRCConnectionReconfigurationComplete* *RRCConnectionReestablishmentComplete* *MeasurementReport*		
SRB1	AM	DCCH	UL-SCH	PUSCH
		ULInformationTransfer (2)		
SRB2	AM	DCCH	UL-SCH	PUSCH

(1): transport of NAS messages only, downstream
(2): transport of NAS messages only, upstream

Table 1.3. *Messages in the RRC protocol*

1.2.2.1. *Information concerning the radioelectric interface*

The information relating to the radioelectric interface is divided between the messages *MasterInformationBlock* (MIB) and *SystemInformationBlock* (SIB). These messages are transmitted periodically, and a change in these data is notified to the mobile by paging.

The MIB message contains the following data:

– the bandwidth of the radioelectric signal for the downstream direction (1.4 MHz, 3, 5, 10, 15, 20 MHz);

– the System Frame Number (SFN);

– the configuration of the physical channel PHICH of the radioelectric interface. The configuration of this physical channel is defined by the operator.

All *SystemInformationBlock* messages, with the exception of the message *SystemInformationBlockType1*, are mapped in a *SystemInformation* message. Each *SystemInformation* message contains *SystemInformationBlock* messages with the same periodicity. The message *SystemInformationBlockType2* must necessarily be mapped in the message *SystemInformation1*.

The message *SystemInformationBlockType1* contains the following data:

– the MCC and MNC of the mobile network;

– the TAC of the TAI;

– the cell identifier;

– the periodicity of the *SystemInformation* messages and the types of *SystemInformationBlock* messages that they contain.

Table 1.4 shows the data transported by the different types of *SystemInformationBlock* message.

SystemInformationBlockType2	Bandwidth in upstream direction Configuration of physical channels
SystemInformationBlockType3	Cell selection parameters
SystemInformationBlockType4	Neighboring EPS cells, same frequency
SystemInformationBlockType5	Neighboring EPS cells, different frequency
SystemInformationBlockType6	Neighboring UMTS cells
SystemInformationBlockType7	Neighboring GSM/GPRS cells
SystemInformationBlockType8	Neighboring CDMA 2000 cells
SystemInformationBlockType9	Identifier of the femtocell eNB
SystemInformationBlockType10 *SystemInformationBlockType11*	Tsunami and earthquake warning

Table 1.4. *SystemInformationBlock messages*

1.2.2.2. *Control of RRC connection*

The different procedures associated with the control of the RRC connection relate to paging, RRC connection setup, activation of security mode, RRC connection reconfiguration, RRC connection re-establishment and RRC connection release.

The message Paging is used by the eNB to alert one or more mobiles in the RRC_IDLE state. It also helps to inform the mobile in RRC_IDLE or RRC_CONNECTED state about a change in the system information or about a notification on the ETWS (Earthquake and Tsunami Warning System) transmitted in the messages *SystemInformationBlockType10* and *SystemInformation BlockType11*.

The message *RRCConnectionRequest* is used by the mobile to request the establishment of an RRC connection.

The message *RRCConnectionSetup* is used by the eNB to configure the SRB1 bearer.

The message *RRCConnectionSetupComplete* is used by the mobile to confirm the setup of the RRC connection. This message can also transport NAS messages.

The message *RRCConnectionReject* is used by the eNB to reject the RRC connection.

Upon receiving the context about the mobile from the MME, the eNB activates security mode for the RRC messages. The messages are only checked for integrity. The encryption of the RRC messages will be effective only if the procedure has been successful.

The message *SecurityModeCommand* is used by the eNB to command the activation of security mode on the radioelectric interface.

The message *SecurityModeComplete* is used by the mobile to confirm the activation of security mode.

The message *SecurityModeFailure* is used by the mobile to indicate that security mode was unable to be activated.

Having initiated the security mode activation procedure, the eNB begins the activation of the DRB. The RRC messages are encrypted and checked for integrity.

The message *RRCConnectionReconfiguration* is used by the eNB to command a modification of the RRC connection. This message may relate to the configuration of the measurements, control of the mobility, configuration of the DRB, etc. This message can also transport NAS messages.

The message *RRCConnectionReconfigurationComplete* is used by the mobile to confirm the reconfiguration of the RRC connection.

The message *RRCConnectionReestablishmentRequest* is used by the mobile to request the re-establishment of the RRC connection.

The message *RRCConnectionReestablishment* is used by the eNB to re-establish the SRB1 bearer.

The message *RRCConnectionReestablishmentComplete* is used by the mobile to confirm the re-establishment of the RRC connection.

The message *RRCConnectionReestablishmentReject* is used by the eNB to indicate that the re-establishment of the RRC connection has been rejected.

The message *RRCConnectionRelease* is used by the eNB to release the RRC connection. The procedure can also be used to redirect the mobile to a different frequency band. In exceptional cases, the mobile can terminate the RRC connection without alerting the eNB.

1.2.2.3. *Measurement report*

The measurements carried out by the mobile must be in line with the configuration indicated in the message *RRCConnectionReconfiguration* transmitted by the eNB. The mobile sends the eNB the measurements in the RRC message *MeasurementReport*.

The configuration of the measurements defines which objects the mobile has to measure:

– the frequency received from the eNB (intra-frequency measurements);

– other frequencies on the EPS network (inter-frequency measurements);

– the frequencies delivered by other mobile networks.

The measurement configuration provides a list of reports containing the criteria giving rise to the sending of a report and the format of those reports. The criteria relate to the events which trigger the report. The format of the report specifies the parameters that need to be included, e.g. the RSRP (Reference Signal Received Power) or the RSRQ (Reference Signal Received Quality).

The measurement configuration determines the identifier of the measurements. Each identifier corresponds to a combination between an object and a report.

1.2.3. *S1-AP protocol*

The S1-AP protocol is the signaling exchanged between the eNB and MME over the S1-MME interface (Figure 1.2). It performs the following functions (Table 1.5):

– activation of the context of the mobile;

– establishment, modification and release of the E-RAB;

– handover management;

– paging. This procedure tells the eNB that the message needs to be broadcast in the cell;

– transport of the NAS signaling exchanged between the mobile and the MME;

– establishment of the S1-MME interface.

1.2.3.1. *Context management*

The context of the mobile has to be established at the level of the eNB and MME so as to transmit the mobile traffic and the NAS signaling. It includes the contexts relating to the default bearer, security, the capacities of the mobile and roaming restrictions. Context setup for the mobile begins with the message INITIAL CONTEXT SETUP REQUEST transmitted by the MME to the eNB. This message follows the reception of the message INITIAL UE MESSAGE.

The MME has to prepare the establishment of the default bearer before receiving the message INITIAL CONTEXT SETUP RESPONSE. This message might contain the cause of the failure to set up the context of the mobile, such as the lack of radioelectric resources.

If the eNB is unable to establish the context of the mobile, it responds with the message INITIAL CONTEXT SETUP FAILURE.

Release of the context of the mobile is done by way of the message UE CONTEXT RELEASE COMMAND transmitted by the MME to the eNB, for instance when the mobile changes cell. This message is acknowledged in return by the response UE CONTEXT RELEASE COMPLETE.

1.2.3.2. *Bearer management*

Establishment and modification of the dedicated bearer E-RAB are initiated by the MME by sending the messages E-RAB SETUP/MODIFY REQUEST. The eNB responds positively or negatively by sending the messages E-RAB SETUP/MODIFY RESPONSE.

Functions	Request	Response
Paging	PAGING	None
Context management	INITIAL CONTEXT SETUP REQUEST	INITIAL CONTEXT SETUP RESPONSE or INITIAL CONTEXT SETUP FAILURE
	UE CONTEXT RELEASE COMMAND	UE CONTEXT RELEASE COMPLETE
Bearer management	E-RAB SETUP/MODIFY REQUEST	E-RAB SETUP/MODIFY RESPONSE
	E-RAB RELEASE COMMAND	E-RAB RELEASE RESPONSE
	E-RAB RELEASE INDICATION	None
Mobility management	HANDOVER REQUIRED	HANDOVER COMMAND
	HANDOVER REQUEST	HANDOVER REQUEST ACKNOWLEDGE or HANDOVER FAILURE
	eNB STATUS TRANSFER	None
	MME STATUS TRANSFER	None
	HANDOVER NOTIFY	None
	PATH SWITCH REQUEST	PATH SWITCH ACKNOWLEDGE or PATH SWITCH FAILURE
S1-MME interface management	S1 SETUP REQUEST	S1 SETUP RESPONSE or S1 SETUP FAILURE
	ENB CONFIGURATION UPDATE	ENB CONFIGURATION UPDATE ACKNOWLEDGE or ENB CONFIGURATION UPDATE FAILURE
	MME CONFIGURATION UPDATE	MME CONFIGURATION UPDATE ACKNOWLEDGE or ENB CONFIGURATION UPDATE FAILURE
	OVERLOAD START	None
	OVERLOAD STOP	None
Transport of NAS signaling	INITIAL UE MESSAGE	None
	DOWNLINK NAS TRANSPORT	None
	UPLINK NAS TRANSPORT	None

Table 1.5. *Messages in the S1-AP protocol*

Release of the dedicated bearer is initiated by the MME by sending the message E-RAB RELEASE COMMAND, or by the eNB by sending an E-RAB RELEASE INDICATION. Upon receiving this message, the MME begins the procedure of release of the dedicated bearer.

1.2.3.3. *Mobility management*

The decision regarding handover based on the S1 interface is made by the source eNB. The phase of handover preparation begins with the sending of the message HANDOVER REQUIRED to the MME. When the reservation of resources by the target eNB is effective, the MME responds with the message HANDOVER COMMAND.

The MME asks the target eNB to reserve the radioelectric resources by way of the message HANDOVER REQUEST. If the operation is successful, the target eNB responds with the message HANDOVER REQUEST ACKNOWLEDGE. This message can contain the elements for construction of a GTP-U tunnel to transfer the received data from the source eNB to the target eNB so they can be transmitted to the mobile. If not, the target eNB responds with the message HANDOVER FAILURE.

The source eNB has to transfer the value of the field SN (Sequence Number) of the protocol PDCP to the target eNB in order to preserve, in the mobile reception, the continuity of the PDCP frame numbering. This operation is done by the transmission of the following messages:

– eNB STATUS TRANSFER from the source eNB to the MME;

– MME STATUS TRANSFER from the MME to the target eNB.

When the execution of the handover has been completed, the target eNB advises the MME of this by way of the message HANDOVER NOTIFY.

The message PATH SWITCH REQUEST is transmitted by the target eNB to the MME for the transfer of the extremity of the GTP-U tunnel corresponding to the source eNB to the target eNB. The MME responds with the message PATH SWITCH ACKNOWLEDGE if the response is positive or with PATH SWITCH FAILURE if not.

1.2.3.4. *S1-MME interface management*

The eNB is in charge of selecting the MME. The eNB therefore takes the initiative to activate the S1-MME interface by transmitting the message S1 SETUP REQUEST, indicating the list of TACs served. The response message from MME, S1 SETUP RESPONSE, contains information relating to the MME such as its

MMEC, the number of the pool to which it belongs and the MNC and MCC. The MME may respond negatively with the message S1 SETUP FAILURE.

Updates to the information about the eNB (or respectively MME) are transmitted by the message ENB CONFIGURATION UPDATE (or respectively MME CONFIGURATION UPDATE). These messages receive a positive response with the messages ENB/MME CONFIGURATION UPDATE ACKNOWLEDGE or a negative one with the messages ENB/MME CONFIGURATION UPDATE FAILURE.

The MME notifies the eNB of the beginning (or respectively the end) of a state of overload by the message OVERLOAD START (or respectively OVERLOAD STOP) so as to avoid being selected for the attachment of a new mobile.

1.2.4. *X2-AP protocol*

The X2-AP protocol is the signaling exchanged between two eNBs over the X2 interface (Figure 1.4). It performs the following functions (Table 1.6):

– mobility management. This function enables the source eNB to transfer the connection of a mobile to the target eNB;

– load management. This function is used by the eNBs to provide an indication of the load of the cells that they serve;

– X2 interface management. This function is used for the activation of the X2 interface, the reconfiguration and re-initialization of the X2 interface.

1.2.4.1. *Mobility management*

The function of mobility management contains the following elementary procedures:

– handover preparation;

– transfer of the state of the field SN of the PDCP protocol;

– deactivation of the context of the mobile;

– handover cancellation.

The procedure of handover preparation is initiated by the source eNB by transmission of the message HANDOVER REQUEST to the target eNB. The target eNB reserves the resources and responds with the message HANDOVER REQUEST ACKNOWLEDGE. This message contains the value of the TEID identifier used by the GTP-U protocol for the traffic transferred by the source eNB to the target eNB. If the resources are unavailable, the target eNB sends back the message HANDOVER PREPARATION FAILURE.

<table>
<tr><th>Functions</th><th>Request</th><th>Response</th></tr>
<tr><td rowspan="2">Load management</td><td>LOAD INFORMATION</td><td>None</td></tr>
<tr><td>RESOURCE STATUS REQUEST</td><td>RESOURCE STATUS RESPONSE
or
RESOURCE STATUS FAILURE</td></tr>
<tr><td>Load management</td><td>RESOURCE STATUS UPDATE</td><td>None</td></tr>
<tr><td rowspan="4">Mobility management</td><td>HANDOVER REQUEST</td><td>HANDOVER REQUEST ACKNOWLEDGE
or
HANDOVER PREPARATION FAILURE</td></tr>
<tr><td>SN STATUS TRANSFER</td><td>None</td></tr>
<tr><td>UE CONTEXT RELEASE</td><td>None</td></tr>
<tr><td>HANDOVER CANCEL</td><td>None</td></tr>
<tr><td rowspan="3">X2 Interface management</td><td>X2 SETUP REQUEST</td><td>X2 SETUP RESPONSE or
X2 SETUP FAILURE</td></tr>
<tr><td>ENB CONFIGURATION UPDATE</td><td>ENB CONFIGURATION UPDATE ACKNOWLEDGE
or
ENB CONFIGURATION UPDATE FAILURE</td></tr>
<tr><td>RESET REQUEST</td><td>RESET RESPONSE</td></tr>
</table>

Table 1.6. *Messages in the X2-AP protocol*

The procedure of transfer of the state of the field SN consists of transferring to the eNB the value of the SN (Sequence Number) of the PDCP protocol with the message SN STATUS TRANSFER. At the source eNB, this message stops the attribution of the SN of the PDCP protocol for the downstream direction.

The procedure for context release of the mobile is initiated by the target eNB by sending the message UE CONTEXT RELEASE to the source eNB. Upon receiving this message, the eNB eliminates the context of the mobile.

The procedure for cancelling the handover is initiated by the source eNB with the message HANDOVER CANCEL. This message causes the target eNB to release the resources on the radioelectric interface.

1.2.4.2. *Load management*

The function of load management includes the following elementary procedures:

– cell load indication;

– initialization of resource status reports;

– resource status reporting.

The procedure for indication of the load of the cell is initiated by either of the eNBs with the message LOAD INFORMATION. This message may contain the following elements of information:

– Interference Overload Indication. This information relates to the interference detected by the eNB, for the upstream direction. The eNB receiving this information has to decrease the level of interference transmitted by the mobile;

– High Interference Indication. This information relates to the interference detected by the eNB, for the upstream direction, indicating which bandwidths are affected. The eNB receiving this information needs to avoid using the said bandwidth, for the upstream direction, for the mobiles located on the periphery of the cell;

– Relative Narrowband Tx Power. This information relates to a decrease in the power transmitted by an eNB. The eNB receiving this information includes it in its traffic management mechanism.

The procedure for initialization of resource status reporting is initiated by either of the eNBs with the message RESOURCE STATUS REQUEST. The eNB receiving this message responds with the message RESOURCE STATUS RESPONSE, which may contain status information for the radioelectric resources, the S1 interface and the load of the eNB. The eNB may respond with RESOURCE STATUS FAILURE if the reports cannot be generated.

The resource status report is then transmitted periodically by the eNB by sending the message RESOURCE STATUS UPDATE.

1.2.4.3. *X2 interface management*

The X2 interface is set up with the intention of exchanging the configuration data necessary for both eNBs to function correctly. One of the eNBs initiates the procedure by indication of the cells served in a message X2 SETUP REQUEST to a candidate eNB. The candidate eNB responds with the message X2 SETUP RESPONSE, also containing the list of cells served. The information communicated may also include the list of nearby cells and the number of antennas for each cell served. The eNB may refuse the establishment of the X2 interface by sending the message X2 SETUP FAILURE in response.

The X2 setup is followed by configuration updating of the eNB if the eNB's configuration changes. The configuration update is initiated by the message ENB CONFIGURATION UPDATE. The distant eNB may respond positively with the message CONFIGURATION UPDATE ACKNOWLEDGE or negatively with the message ENB CONFIGURATION UPDATE FAILURE.

The reset of the X2 interface is intended to align the resources of the eNBs in the case of an unexpected breakdown. The procedure is initiated by the message RESET REQUEST. The receiving eNB responds with the message RESET RESPONSE. The procedure does not affect the data exchanged during the X2 setup or the eNB configuration update.

1.2.5. *GTPv2-C protocol*

GTP tunnels are used between two entities. Such tunnels enable the traffic or signaling data to be compartmentalized. GTP-U traffic tunnels are constructed on the interfaces S1-U, S5 and X2. GTP-C signaling tunnels are created on the S5, S11 and S10 interfaces.

The tunnel is identified by the parameters TEID, the IP addresses and the UDP port numbers. The entity receiving the traffic or signaling data determines the value of the parameter TEID which the sending entity has to use.

The values of the parameter TEID of the GTP-U protocol are exchanged via the GTPv2-C, S1-AP and X2-AP protocols. This parameter is used for data flows belonging to the same QCI.

The TEID used for the signaling exchanged over the S5 interface is unique. The same parameter is used for all signaling messages relating to the activation of the various S5 bearers for the different mobiles.

The TEID used for the signaling exchanged over the S10 and S11 interfaces is unique for each mobile. The same parameter is used for all signaling messages relating to the establishment of the various S1-U bearers for the same mobile.

Table 1.7 recaps the messages making up the GTPv2-C protocol.

<table>
<tr><th>Type of messages</th><th>Request</th><th>Response</th></tr>
<tr><td rowspan="5">Mobility management</td><td>FORWARD RELOCATION REQUEST</td><td>FORWARD RELOCATION RESPONSE</td></tr>
<tr><td>FORWARD RELOCATION NOTIFICATION</td><td>FORWARD RELOCATION ACKNOWLEDGE</td></tr>
<tr><td>FORWARD ACCESS CONTEXT NOTIFICATION</td><td>FORWARD ACCESS CONTEXT ACKNOWLEDGE</td></tr>
<tr><td>CONTEXT REQUEST</td><td>CONTEXT RESPONE</td></tr>
<tr><td colspan="2">CONTEXT ACKNOWLEDGE</td></tr>
<tr><td rowspan="3">Bearer management</td><td>CREATE/DELETE SESSION REQUEST</td><td>CREATE/DELETE SESSION RESPONSE</td></tr>
<tr><td>CREATE/MODIFY/DELETE BEARER REQUEST</td><td>CREATE/MODIFY/DELETE BEARER RESPONSE</td></tr>
<tr><td>DOWNLINK DATA NOTIFICATION</td><td>DOWNLINK DATA NOTIFICATION ACKNOWLEDGE
or
DOWNLINK DATA NOTIFICATION FAILURE INDICATION</td></tr>
<tr><td>Bearer management</td><td>CREATE/DELETE INDIRECT DATA FORWARDING TUNNEL REQUEST</td><td>CREATE/DELETE INDIRECT DATA FORWARDING TUNNEL RESPONSE</td></tr>
</table>

Table 1.7. *Messages in the GTPv2-C protocol*

1.2.5.1. *Bearer management*

The signaling bearer unique to a mobile is created by the message CREATE SESSION REQUEST. It is reinforced by the use of a TEID. The message is transmitted:

– by the MME to the SGW, over the S11 interface;

– by the target SGW for the PGW over the S5 interface.

The request is transmitted when any of the following procedures are initiated:

– attachment of a mobile;

– traffic request from the mobile;

– updating of the TAC, leading to a switch of the SGW;

– cell handover, with switch of the SGW.

The SGW (or respectively PGW) responds to the MME (or respectively SGW) with the message CREATE SESSION RESPONSE.

The signaling bearer is deactivated by the exchange of the messages DELETE SESSION REQUEST/RESPONSE. The procedure is triggered when the mobile detaches, when the traffic is released, when the TAC changes, leading to a modification of the SGW, or when handover occurs, with a switch of the SGW.

Similarly, the traffic-dedicated bearer specific to a mobile is created, possibly modified and deleted by the exchange of the following messages:

– CREATE/MODIFY/DELETE BEARER REQUEST;

– CREATE/MODIFY/DELETE BEARER RESPONSE.

The message DOWNLINK DATA NOTIFICATION is sent by the SGW to the MME, over the S11 interface. The procedure follows the SGW's receipt of data from the PDW, with the mobile in ECM-IDLE mode. Upon receiving this message, the MME sends the paging message to the mobile.

The MME may respond with the message DOWNLINK DATA NOTIFICATION ACKNOWLEDGE, indicating whether or not the request is accepted, or DOWNLINK DATA NOTIFICATION FAILURE INDICATION if the mobile does not respond to the paging or if the mobile service request is rejected.

The messages CREATE INDIRECT DATA FORWARDING TUNNEL REQUEST/RESPONSE create a specific traffic bearer when cell handover occurs. This bearer channels the data traffic received by the source eNB to the SGW to then be re-transmitted to the mobile via the target eNB.

1.2.5.2. *Mobility management*

Mobility management messages are exchanged between the source and target MMEs, when the handover of the mobile imposes a switch of MME.

The source MME sends the target MME the message FORWARD RELOCATION REQUEST, containing the context of the mobile. The target MME responds with the message FORWARD RELOCATION RESPONSE when the resources needed for the handover have been reserved. The response contains the values of the TEIDs, which will enable the source SGW to redirect the traffic to the target SGW during handover. Upon receiving this message, the source MME sends the source eNB the command to initiate handover.

The source MME sends the target MME the message FORWARD ACCESS CONTEXT NOTIFICATION to furnish it with the elements of the context of the E-RAB bearer, such as the PDCP sequence number.

The target MME sends the source MME the message FORWARD RELOCATION NOTIFICATION to indicate that the handover procedure is complete.

The new MME sends the message CONTEXT REQUEST to the former one in the procedure of TAI updating, to retrieve information about the context of the mobile. The former MME provides this information in the message CONTEXT RESPONSE, which may contain a positive or negative response. The new entity acknowledges this previous message with the message CONTEXT ACKNOWLEDGE.

1.3. Procedures

1.3.1. *Attachment procedure*

The attachment procedure comprises the following stages:

– mutual authentication between the mobile and the MME;

– location of the mobile by the MME;

– establishment of a default bearer. If the telephone service is borne by the EPS network, a default bearer (QCI = 5) is created to transport the telephone signaling exchanged between the mobile and the IMS network;

– assignment of a temporary private identity (GUTI).

1.3.1.1. *Registration*

The registration procedure is described in Figure 1.6.

The procedure for the registration of the mobile with the MME is preceded by the procedure of connection of the mobile to the eNB.[3]

The registration procedure is triggered by the mobile when it sends the MME the messages EMM ATTACH REQUEST and ESM PDN CONNECTIVITY REQUEST containing its IMSI (International Mobile Subscriber Identity) or GUTI,

3 A description of the procedure for connection to the eNB is given in Chapter 2, on the LTE interface.

if known. The selection of the MME attributed to the mobile is performed by the eNB, which transfers the EMM and ESM messages.

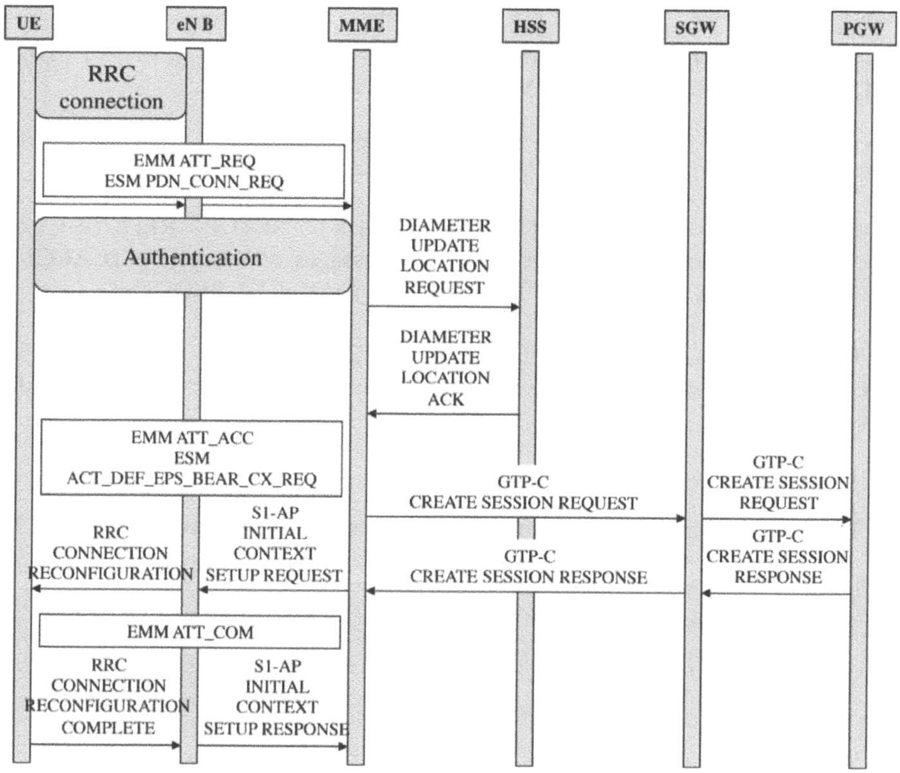

Figure 1.6. *Registration procedure*

The EMM and ESM messages are transmitted:

– by the message *RRCConnectionSetupComplete* over the LTE-Uu radioelectric interface;

– by the message S1-AP INITIAL UE MESSAGE over the S1-MME interface.

Following the procedure of mutual authentication, the MME updates the location of the mobile in the HSS database, via the message DIAMETER UPDATE LOCATION REQUEST.

In return, the HSS provides the MME with the profile of the mobile via the message DIAMETER UPDATE LOCATION ACK.

The default bearer is established by the following signaling exchanges:

– over the S11 and S5 interfaces, the messages GTP-C CREATE SESSION REQUEST and CREATE SESSION RESPONSE;

– over the S1-MME interface, the messages S1-AP INITIAL CONTEXT SETUP REQUEST and INITIAL CONTEXT SETUP RESPONSE;

– over the LTE-Uu radioelectric interface, the RRC messages *ConnectionReconfiguration* and *RRCConnectionReconfigurationComplete*.

The messages S1-AP INITIAL CONTEXT SETUP REQUEST and *RRCConnectionReconfiguration* transport the messages EMM ATTACH ACCEPT and ESM ACTIVATE DEFAULT EPS BEARER CONTEXT REQUEST.

The messages S1-AP INITIAL CONTEXT SETUP RESPONSE and *RRCConnectionReconfigurationComplete* transport the message EMM ATTACH COMPLETE.

1.3.1.2. *Authentication and security procedure*

The procedure for authentication and security activation is shown in Figure 1.7.

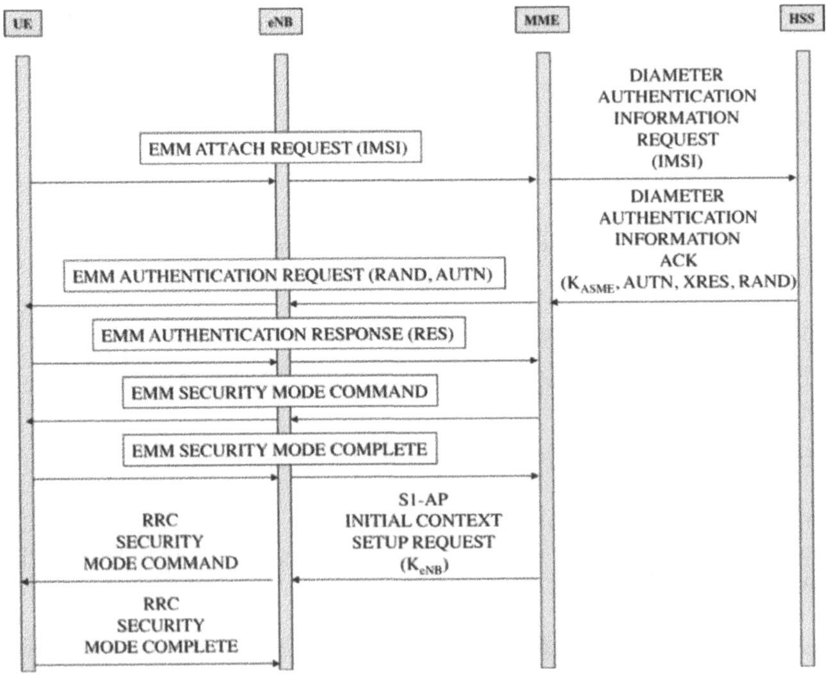

Figure 1.7. *Authentication and security mode procedure*

Upon receipt of the message EMM ATTACH REQUEST, an exchange of messages DIAMETER AUTHENTICATION INFORMATION REQUEST/ACK enables the MME to recover the following data from the HSS:

– a random number RAND;

– the authentication code XRES, computed on the basis of the RAND and the secret key Ki of the mobile;

– the network authentication AUTN;

– the key K_{ASME}, derived from the IK and CK keys, which themselves are derived from the Ki.

On the basis of the K_{ASME}, the MME derives the following keys:

– CK_{NAS}, used to encrypt the NAS messages;

– IK_{NAS}, used to verify the integrity of the NAS messages;

– Ke_{NB}, transferred to the eNB.

The MME sends the mobile the message EMM AUTHENTICATION REQUEST containing the RAND and the AUTN.

The mobile uses its secret Ki stored in the USIM (Universal Services Identity Module) on its UICC (Universal Integrated Circuit Card), and the RAND received, to locally compute its own authentication code RES and that of the network, which it compares to the AUTN value received. If the two values are identical, the network is authenticated.

The mobile also derives the keys CK_{NAS}, IK_{NAS} and Ke_{NB} from the Ki and the RAND received.

The mobile responds to the MME with the message EMM AUTHENTICATION RESPONSE, containing the authentication code RES. The MME compares the received values RES from the mobile and XRES from the HSS. If the values are identical, the mobile is authenticated.

The security mode for NAS signaling is activated by the exchange of the messages EMM SECURITY MODE COMMAND/COMPLETE.

The above EMM messages are transported:

– by the S1-AP messages DOWNLINK NAS TRANSPORT and RRC *DLInformationTransfer* for the downstream direction;

– by the messages RRC *ULInformationTransfer* and S1-AP UPLINK NAS TRANSPORT for the upstream direction.

The MME sends the eNB the key K_{eNB} in the message S1-AP INITIAL CONTEXT SETUP REQUEST.

Using the K_{eNB}, the eNB and mobile can derive the following keys:

– $CK_{eNB\text{-}RRC}$, used for the encryption of the RRC messages;

– $K_{eNB\text{-}RRC}$, used to check the integrity of the RRC messages;

– $CK_{eNB\text{-}UP}$, used to encrypt the data traffic.

Security mode for the LTE-Uu radioelectric interface is activated by the exchange of the RRC messages *SecurityModeCommand* and *SecurityMode Complete*.

1.3.2. *Location update*

The procedure of location updating is illustrated in Figure 1.8.

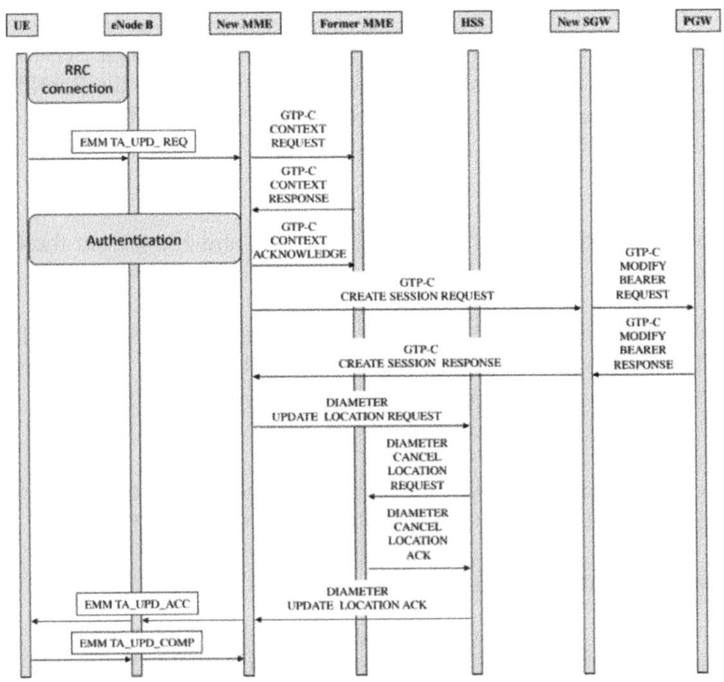

Figure 1.8. *Tracking area update procedure*

The procedure of registration of the mobile with the MME is preceded by the procedure of connection of the mobile to the eNB.

The mobile decides to update its location when it enters into a new TAI or when the hold-down timer has expired. The updating of the location includes the following operations:

– transfer of the context of the mobile from the former MME to the new one, if the new TAC belongs to a different pool;

– mutual authentication between the mobile and the new MME;

– updating of the bearer between the new SGW and the PGW;

– updating of the HSS with the identity of the new MME.

The mobile initiates the procedure of location update by sending the message EMM TRACKING AREA UPDATE REQUEST to the new MME.

The message EMM TRACKING AREA UPDATE REQUEST is transported:

– by the message *RRCConnectionSetupComplete* over the LTE-Uu radioelectric interface, between the mobile and the eNB;

– by the message S1-AP INITIAL UE MESSAGE over the S1-MME interface between the eNB and MME.

The new MME recovers the identity of the former one from the mobile GUTI, and contacts it, transmitting the message GTP-C CONTEXT REQUEST. The former MME sends back the context of the mobile in the message GTP-C CONTEXT RESPONSE.

The authentication procedure is performed between the mobile and the new MME, which then acknowledges transfer from the former MME by sending it the message GTP-C CONTEXT ACKNOWLEDGE.

The new MME creates a session with a new SGW by an exchange of signaling GTP-C CREATE SESSION REQUEST/RESPONSE. This message contains the IP address of the PGW, retrieved from the context of the mobile. The new SGW modifies the S5 bearer with the PGW by an exchange of signaling GTP-C MODIFY BEARER REQUEST/RESPONSE.

The new MME notifies the HSS of the updated location of the mobile by exchange of messages DIAMETER UPDATE LOCATION REQUEST/ACK.

For its part, the HSS cancels the data about the mobile in the former MME by exchanging the messages DIAMETER CANCEL LOCATION REQUEST/ACK. Upon receiving this message, the former MME deletes the context of the mobile. It performs the same operation on the SGW, deleting the session established by the exchange of the messages GTP-C DELETE SESSION REQUEST/RESPONSE.

The new entity can then respond to the mobile by sending the message EMM TRACKING AREA UPDATE ACCEPT containing a new GUTI. This message is acknowledged in return by the mobile with the message EMM TRACKING AREA UPDATE COMPLETE.

1.3.3. *Bearer activation*

The establishment of a dedicated bearer for voice data (QCI = 1) is coupled with the re-establishment of the default bearer attributed to the telephone signaling (QCI = 5).

The re-establishment of the default bearer is triggered by a service request. The procedure is initiated by the mobile in idle mode (with an outgoing call) or by the network (with an incoming call) to establish the DRB and S1 bearers.

1.3.3.1. *Re-establishment of the default bearer with an outgoing call*

The procedure for re-establishment of the default bearer in the case of an outgoing call is that shown in Figure 1.9.

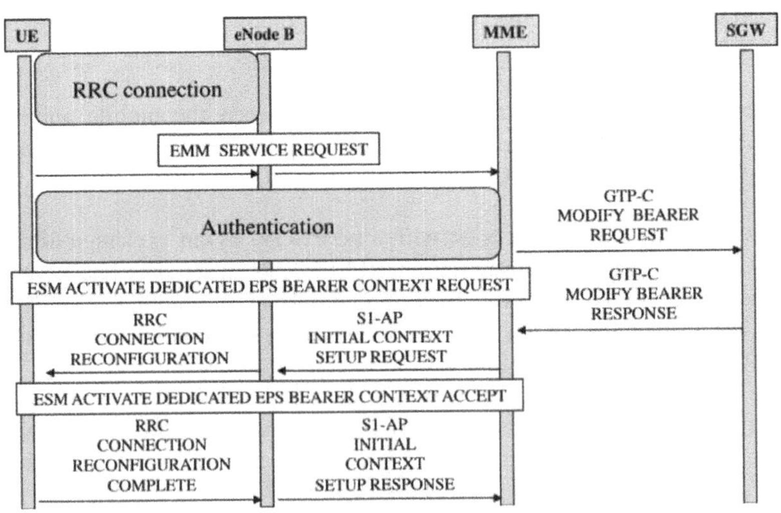

Figure 1.9. *Re-establishment of the default bearer with an outgoing call*

When making an outgoing call, the mobile initiates the service request procedure by activating the RRC connection and then sending a message EMM SERVICE REQUEST.

After authentication, the MME triggers the activation of the bearer on the S1-U interface by exchanging the messages GTP-C MODIFY BEARER REQUEST/RESPONSE with the SGW. The MME performs the same operation on the eNB, exchanging the messages S1-AP INITIAL CONTEXT SETUP REQUEST/RESPONSE. It should be noted that the S5 bearer remains active.

The eNB establishes the bearer on the radioelectric interface by exchange of the signaling *RRCReconfiguration* and *RRCReconfigurationComplete.*

1.3.3.2. *Re-establishment of the default bearer with an incoming call*

The procedure of re-establishment of the default bearer in the case of an incoming call is shown in Figure 1.10.

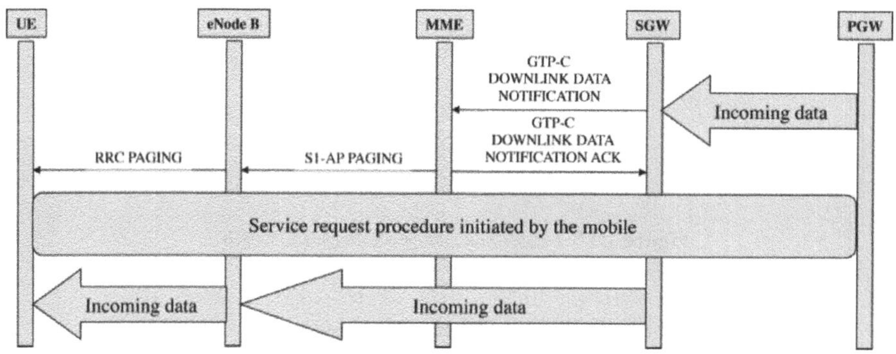

Figure 1.10. *Re-activation of the default bearer with an incoming call*

With an incoming call, because the S5 bearer is active, the SGW receives the telephone signaling data from the PGW, but the LTE-Uu and S1 bearers are inactive, as the mobile is in idle mode. The SGW generates the message GTP-C DOWNLINK DATA NOTIFICATION, sent to the MME.

The MME initiates the paging procedure by sending the eNB the message S1-AP PAGING, containing the S-TMSI (shortened temporary mobile subscriber) for the mobile.

The eNB transmits the RRC message *Paging* over the Uu interface. Upon receiving this message, the mobile initiates the service request procedure.

When the DRB and S1 bearers are established, the SGW transfers the stored incoming data to the eNB, which in turn sends them to the mobile.

1.3.3.3. *Establishment of the dedicated bearer*

The procedure for establishment of the dedicated bearer for voice data is shown in Figure 1.11.

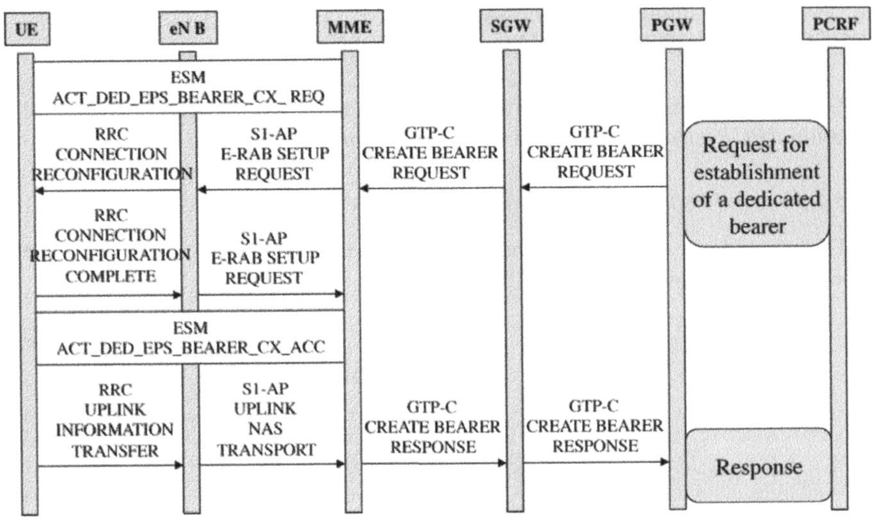

Figure 1.11. *Establishment of the dedicated bearer*

The establishment of the dedicated bearer is triggered by the IMS network, on the basis of the analysis of the telephone signaling exchanged between the terminals wishing to establish telephone communication.

The link between the IMS and EPS networks is ensured by the PCRF. This entity sends the PGW the characteristics of the voice-data dedicated bearer that needs to be activated. This dedicated bearer is coupled with the default bearer assigned to the telephone signaling, in the sense that the endings are identical for the two types of bearer.

The messages exchanged for the establishment of the dedicated bearer are as follows:

– for establishment of the S5 bearer, the GTP-C messages exchanged between the PGW and SGW (CREATE BEARER REQUEST/RESPONSE);

– for establishment of the S1 bearer, the GTP-C messages exchanged between the SGW and MME (CREATE BEARER REQUEST/RESPONSE) and the S1-AP messages exchanged between the MME and eNB (E-RAB SETUP REQUEST/RESPONSE);

– for establishment of the DRB, the RRC messages (*ConnectionReconfiguration* and *ConnectionReconfigurationComplete*) exchanged between the mobile and the eNB.

The establishment of the dedicated bearer is dependent upon the mobile's acceptance of it, in the wake of the exchange of ESM messages between the mobile and the MME:

– the request ESM ACTIVATE DEDICATED EPS BEARER CONTEXT REQUEST is sent by the MME;

– the response ESM ACTIVATE DEDICATED EPS BEARER CONTEXT ACCEPT is sent back by the mobile.

1.3.4. *Handover procedure*

The procedure of handover takes place in two phases:

– the preparation phase, corresponding to the handover decision and to the reservation of the resources;

– the execution phase, corresponding to the handover itself, to the connection of the mobile to the target eNB and the release of the old resources.

1.3.4.1. *Handover based on the X2 interface*

This scenario refers to the activation of the X2 interface between two eNBs – the source and the target. Two procedures are possible, depending on whether or not the MME and SGW entities need to be relocated. The procedure without relocation is shown in Figure 1.12.

Upon receiving RRC messages *MeasurementReport*, the source eNB takes the decision to perform a cell handover and sends the target eNB the message X2-AP HANDOVER REQUEST.

On receiving the message X2-AP HANDOVER REQUEST ACK, the eNB triggers handover by sending the mobile the message *RRCConnection Reconfiguration*.

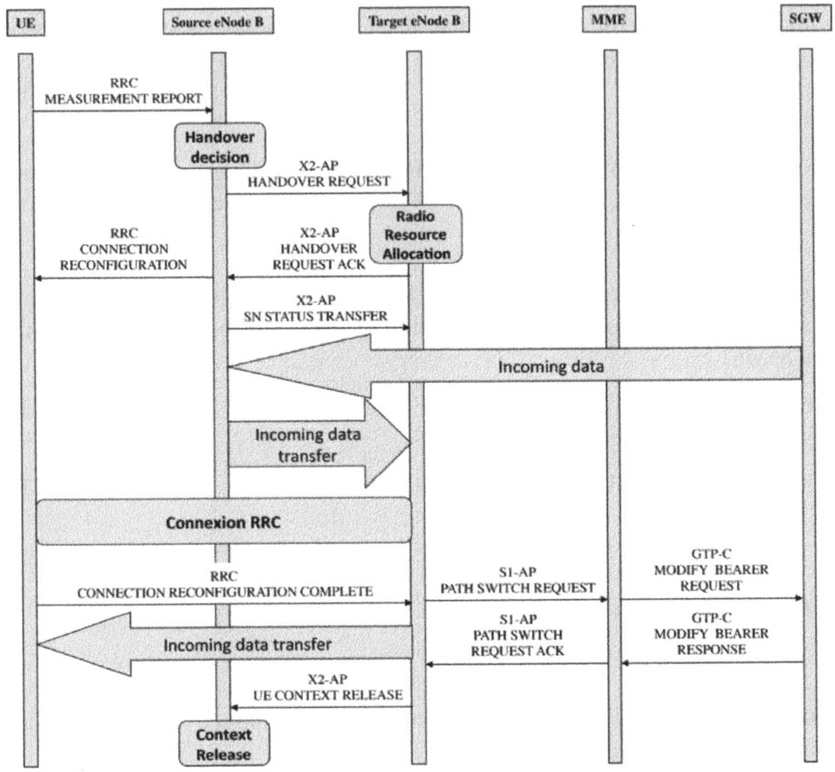

Figure 1.12. *Handover based on the X2 interface, without relocation*

The source eNB also sends the target eNB the PDCP sequence numbers in the message X2-AP SN STATUS TRANSFER, and then the incoming data which have not been acknowledged by the mobile. The target eNB will store these data until the mobile is able to receive them.

When the RRC connection is established between the mobile and the target eNB, the mobile sends a message *RRCConnectionReconfigurationComplete*, which triggers the transfer of the incoming data to the mobile and the modification of the S1-U bearer, by exchange of the following messages:

– S1-AP PATH SWITH REQUEST to the MME;

– GTP-C UPDATE BEARER REQUEST to the SGW.

When the S1-U bearer is available, the target eNB alerts the source eNB by the message X2-AP UE CONTEXT RELEASE so that it will release the context associated with the mobile.

1.3.4.2. *Handover based on the S1 interface, without relocation*

The procedure of handover based on the S1 interface occurs when the X2 interface is inactive. It may result in the relocation of the MME and SGW. The procedure without relocation is illustrated by Figure 1.13.

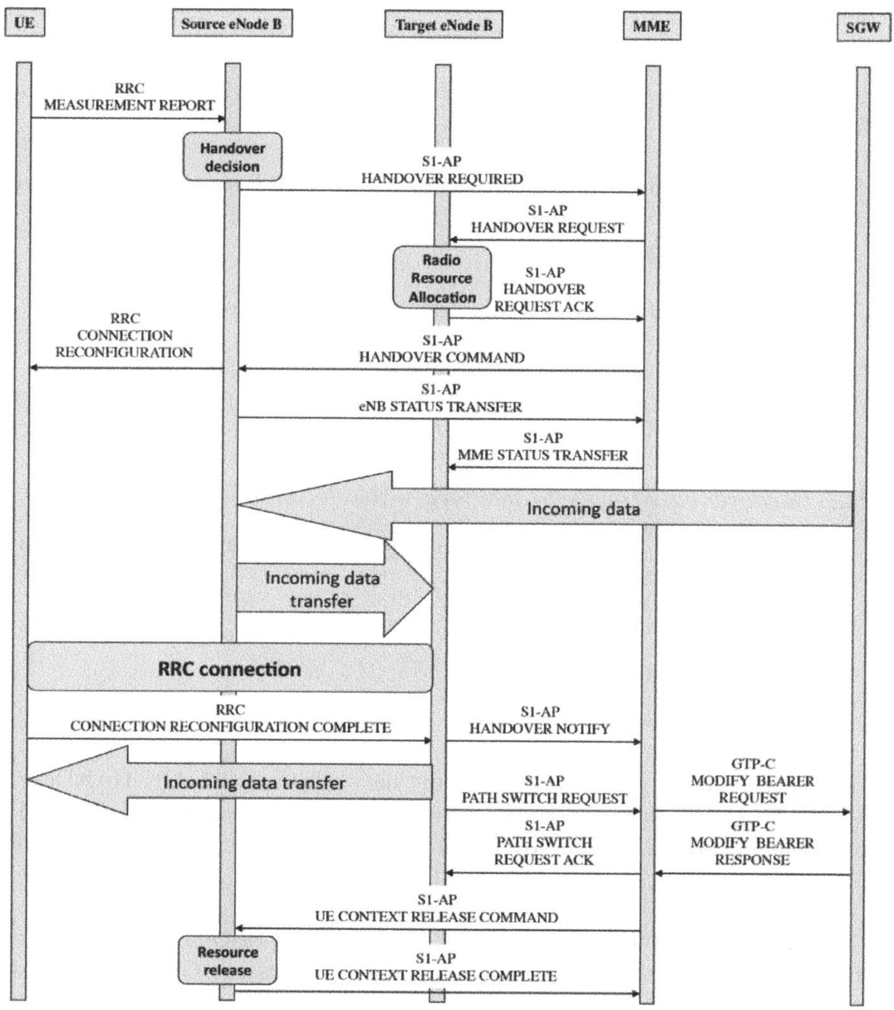

Figure 1.13. *Handover based on the S1 interface, without relocation*

The MME is no longer transparent to the handover mechanism, and plays the role of a signaling relay for the handover command between the source and target eNBs:

– upon receiving the message S1-AP HANDOVER REQUIRED from the source eNB, the MME generates and sends the message S1-AP HANDOVER REQUEST to the target eNB;

– upon receiving the message S1-AP HANDOVER REQUEST ACK from the target eNB, the MME sends the S1-AP HANDOVER COMMAND to the source eNB.

Similarly, the sequence number transfer takes place by way of the messages S1-AP ENB STATUS TRANSFER to the MME and S1-AP MME STATUS TRANSFER to the target eNB.

The modification of the S1-U bearer occurs by the same procedure as before.

The release of the context of the mobile at the source eNB is triggered by the MME, by sending the message S1-AP UE CONTEXT RELEASE COMMAND to the source eNB.

1.3.4.3. *Handover based on the S1 interface, with relocation*

The procedure of handover based on the S1 interface with relocation is illustrated in the following figures:

– Figure 1.14 for handover preparation;

– Figure 1.15 for handover execution;

– Figure 1.16 for context release.

The preparation phase begins with the message S1-AP HANDOVER REQUIRED, sent by the source eNB which has taken the handover decision. This information is relayed by the former MME to the new in the message GTP-C FORWARD RELOCATION REQUEST (Figure 1.14).

The information in the message S1-AP HANDOVER REQUEST ACK received by the new MME is relayed to the former one by the message GTP-C FORWARD RELOCATION RESPONSE (Figure 1.14).

The former and new MMEs create a specific tunnel between the former and new SGWs, by the messages CREATE INDIRECT DATA FORWARDING TUNNEL

REQUEST/RESPONSE. This tunnel is used to transfer the traffic data received by the source eNB (incoming data) to the target eNB during handover of the mobile (Figure 1.14).

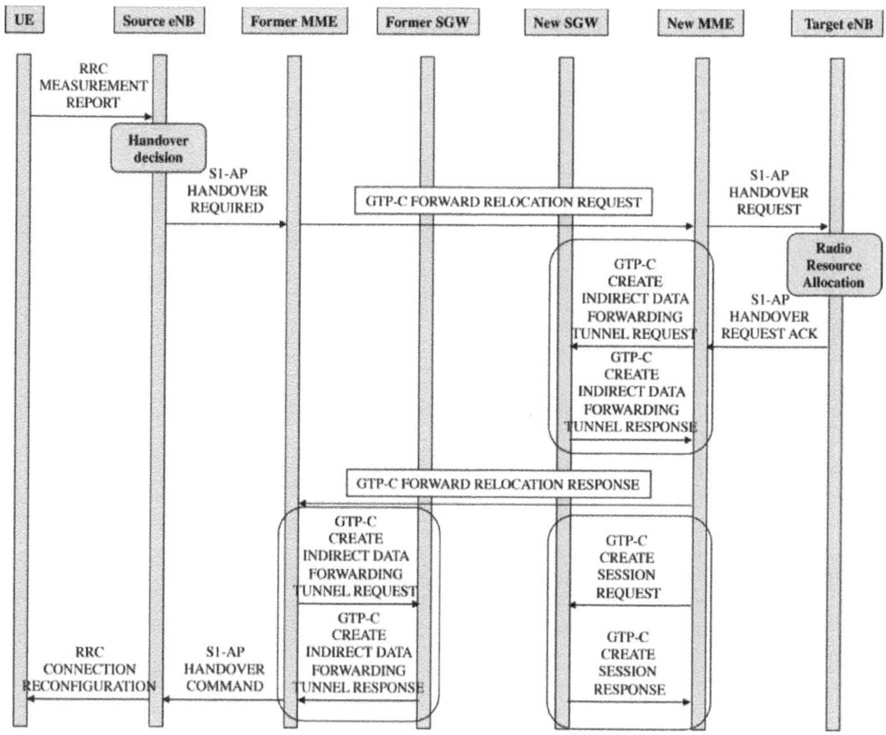

Figure 1.14. *Handover based on the S1 interface, with relocation – preparation*

The new MME creates a session with the new SGW to activate the S1-U bearer, by way of the messages GTP-C CREATE SESSION REQUEST/RESPONSE (Figure 1.14).

When the handover command has been sent to the mobile, the source eNB sends the message S1-AP eNB STATUS TRANSFER to the former MME. This information is relayed to the new MME by the message GTP-C FORWARD ACCESS CONTEXT NOTIFICATION, and then to the target eNB by the message S1-AP MME STATUS TRANSFER (Figure 1.15).

When the mobile is connected to the target eNB, the new eNB sends the message S1-AP HANDOVER NOTIFY to the new MME. This information is relayed to the former MME by the message GTP-C FORWARD RELOCATION NOTIFICATION (Figure 1.15).

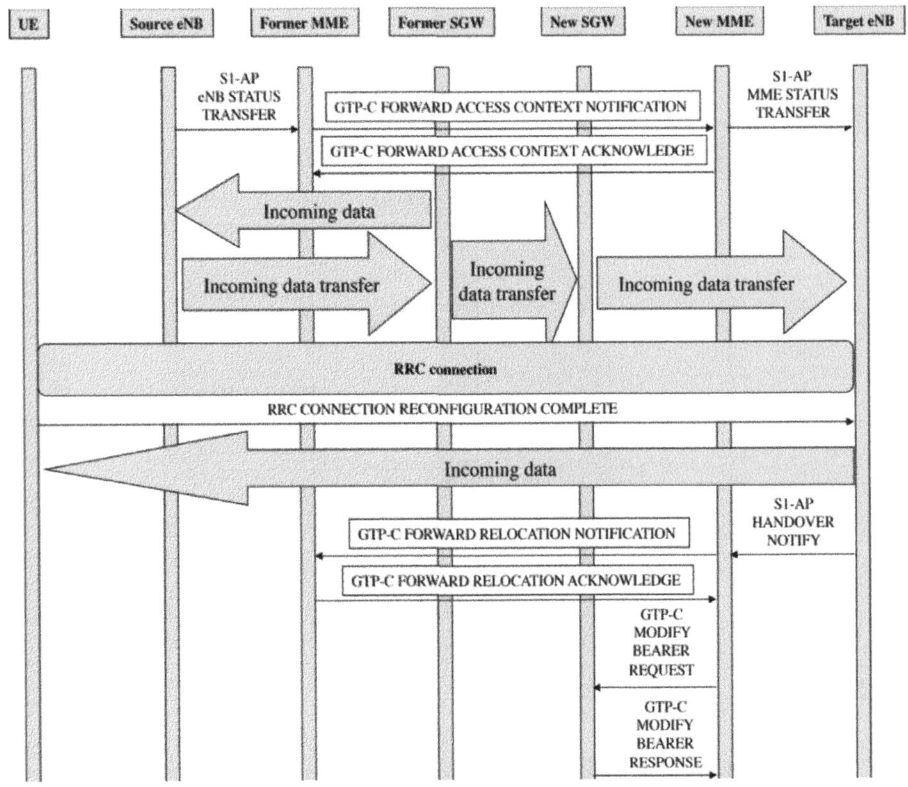

Figure 1.15. *Handover based on the S1 interface, with relocation – execution*

The last phase is the release of the following contexts and bearers:

– the context of the mobile in the source eNB, by the former MME's sending of the message S1-AP UE CONTEXT RELEASE (Figure 1.16);

– the S1-U bearer between the former SGW and the source eNB, by the exchange of the messages GTP-C DELETE SESSION REQUEST/RESPONSE between the former MME and former SGW (Figure 1.16);

– the specific tunnel created between the former and the new SGW, by the exchange of the messages DELETE INDIRECT DATA FORWARDING TUNNEL REQUEST/RESPONSE (Figure 1.16).

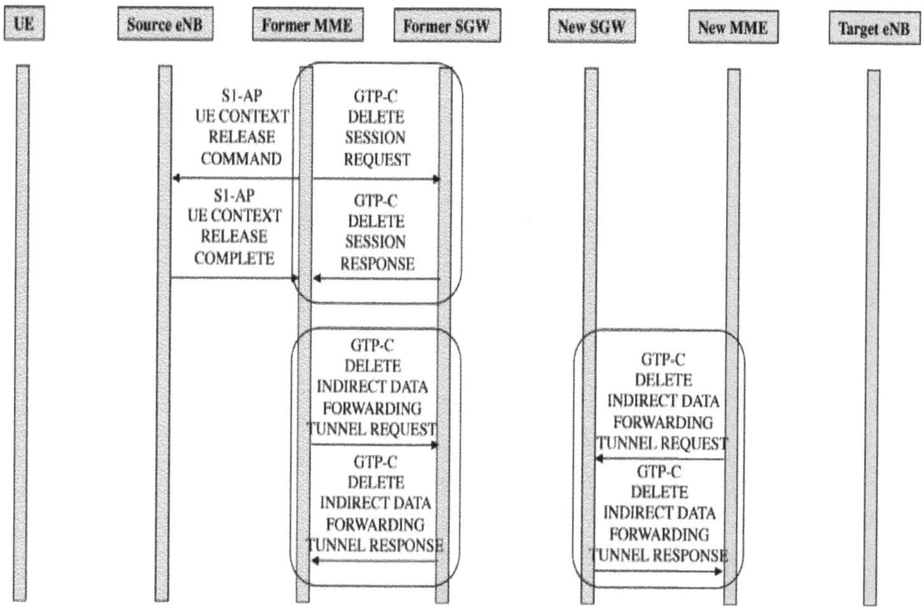

Figure 1.16. *Handover based on the S1 interface, with relocation – release*

Chapter 2

The LTE Interface

2.1. Structure of the radioelectric interface

At the LTE-Uu radioelectric interface, between the mobile UE (User Equipment) and the eNB (evolved Node B), the DRB (Data Radio Bearer) traffic data, corresponding to an IP packet, and the SRB (Signaling Radio Bearer) data, corresponding to an RRC (Radio Resource Control) message, are encapsulated by the data link layer, which is broken down into three sub-layers (Figure 2.1):

– PDCP (Packet Data Convergence Protocol);

– RLC (Radio Link Control);

– MAC (Medium Access Control).

Three levels of channels are defined (Figure 2.1):

– the logic channel defines the type of information. It corresponds to the data structure at the interface between the RLC and MAC sub-layers;

– the transport channel defines the data format. It corresponds to the data structure at the interface between the MAC sublayer and the physical layer. The data are transmitted to the physical layer in the form of codewords. One or two blocks are delivered for each TTI (Transmission Time Interval) of duration of 1ms;

– the physical channel defines the resource attributed to the transport channel and its positioning in the frame. Certain physical channels are used solely for the control of the physical layer and do not include transport channels.

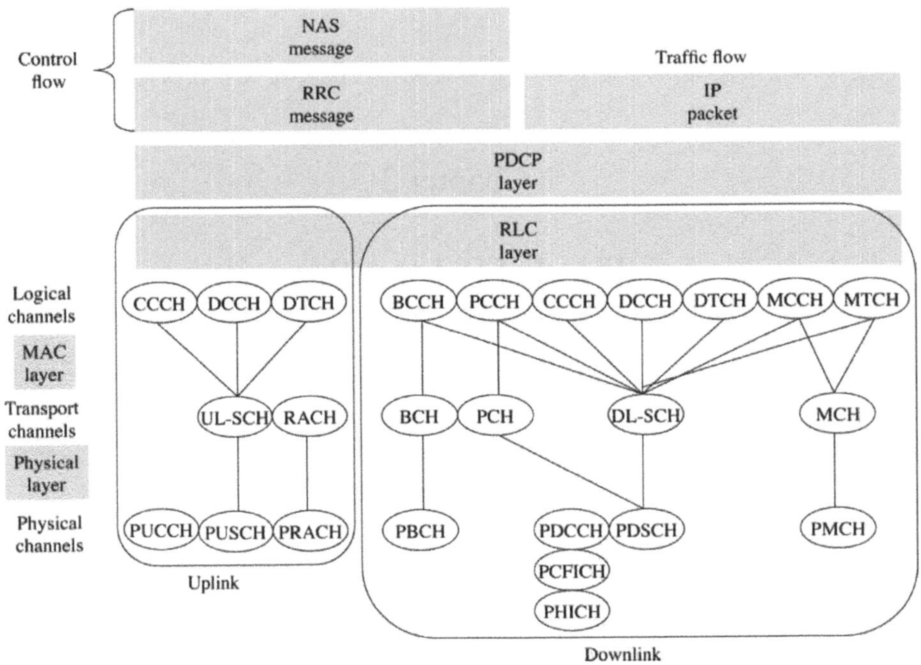

Figure 2.1. *Structure of the radio electric interface*

2.2. Data link layer

2.2.1. *PDCP protocol*

The protocol PDCP is responsible for the following functions:

– Robust Header Compression (ROHC);

– security of data traffic (confidentiality) and RRC signaling (integrity and confidentiality);

– sequential delivery of PDCP frames of traffic when handover occurs.

The protocol PDCP is used for the RRC signaling data mapped in the logic channel DCCH and the traffic data mapped in the logic channel DTCH. The RRC signaling data mapped in the other logic channels do not use PDCP.

The mobile can simultaneously activate several instances of PDCP – two instances for the RRC signaling data and one instance for each active DRB. In the case of telephone communication, two PDCP instances are active – one for telephone signaling and the other for voice data. These two instances correspond to two DRBs.

PDCP defines headers to encapsulate the RRC signaling data and traffic data. PDCP also specifies control messages associated with the traffic data (Figure 2.2).

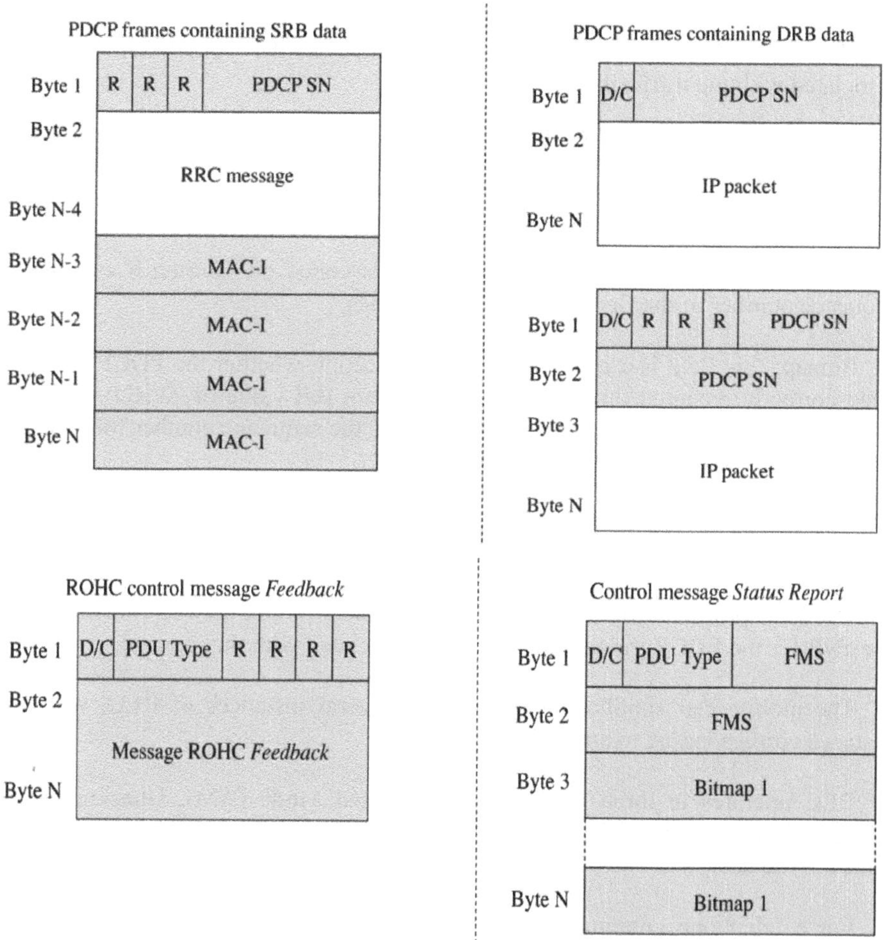

Figure 2.2. *Structure of the PDCP frame*

PDCP SN (Sequence Number): this field is coded on 5 bits for the RRC signaling data and on 7 or 12 bits for the traffic data. It indicates the sequence number of the PDCP frame. This sequence number can be used to recover the PDCP blocks that are lost during handover.

MAC-I (Message Authentication Code for Integrity): this field is coded on 4 bytes. It contains the code used to check the integrity of the PDCP frame containing RRC signaling data.

D/C (Data/Control): this bit indicates whether, or not, the frame contains traffic data (bit value of ONE) or control messages specific to PDCP (bit value of ZERO).

PDU Type: this field is coded on 3 bits. It indicates the type of control message associated with the traffic data:

– the *Status Report* is used in the procedure of re-establishment following a cell handover. It shows a list of PDCP frame numbers not received during the handover;

– the message ROHC *Feedback* is linked to the ROHC mechanism.

FMS (First Missing PDCP SN): this field is coded on 12 bits. It contains the sequence number of the first missing PDCP frames.

Bitmap: this field is a collection of bits indicating whether the PDCP frame has been correctly received (bit value of ONE) or not (bit value of ZERO). The Most Significant Bit (MSB) in the first byte represents the sequence number following the value of the FMS.

2.2.2. *RLC protocol*

The RLC protocol offers control of the radioelectric link between the mobile and the eNB for the PDCP frames and signaling associated with that control.

The mobile can simultaneously activate several instances of RLC, with each instance corresponding to an instance of PDCP.

RLC operates in three modes: Acknowledged Mode (AM), Unacknowledged Mode (UM) and Transparent Mode (TM). For Transparent Mode, no header is added to the data.

For a telephone communication, the voice data is channeled in UM so as to optimize the delay. If an IP packet is lost, the codec employs a Packet-Loss Concealment (PLC) mechanism.

RLC performs the following operations:

– re-transmission in the case of error, with the mechanism ARQ (Automatic Repeat reQuest), only for AM;

– concatenation, segmentation and re-assembling of the PDCP frames, in AM or UM;

– re-segmentation of PDCP data, in AM, when re-transmitting the RLC frame;

– re-sequencing of received data, in AM or UM;

– duplicate data detection, in AM or UM.

2.2.2.1. *Logic channels*

The BCCH (Broadcast Control Channel) is a unidirectional control channel, used only for downlink to carry system information. This channel is mapped on the transport channel BCH for the RRC message *MasterInformationBlock* and on the transport channel DL-SCH for the messages RRC *SystemInformationBlock*. This channel uses transparent mode.

The PCCH (Paging Control Channel) is a unidirectional channel, used only for downlink to carry RRC paging messages. This channel uses TM.

The CCCH (Common Control Channel) is a bidirectional channel, used to transmit the first RRC signaling messages when the mobile connects to the eNB. This channel uses TM.

The DCCH (Dedicated Control Channel) is a bidirectional channel, used to transmit RRC messages when the mobile is connected to the eNB. This channel uses AM.

The MCCH (Multicast Control Channel) is a unidirectional channel used to transmit control information in conjunction with a multicast traffic. This channel is mapped either on the MCH transport channel if the multicast traffic is being transmitted only within the cell, or on the DL-SCH transport channel if the multicast traffic is being transmitted over several cells.

The DTCH (Dedicated Traffic Channel) is a bidirectional dedicated channel, used to transmit unicast IP packets. This channel uses AM or UM depending on the type of service.

The MTCH (Multicast Traffic Channel) is a unidirectional channel used to transmit multicast IP packets to the mobile. As with the MCCH, this channel is mapped either on the MCH or on the DL-SCH.

2.2.2.2. *Protocol structure*

RLC defines a header for UM, a header for AM and a control message.

In UM, the structure of the header depends on the size of the SN and on the number of concatenated PDCP frames. Figure 2.3 shows the RLC header for the following two cases:

– a 5-bit SN, without concatenation of PDCP frames;

– a 5-bit SN, with concatenation of three PDCP frames, with no segmentation.

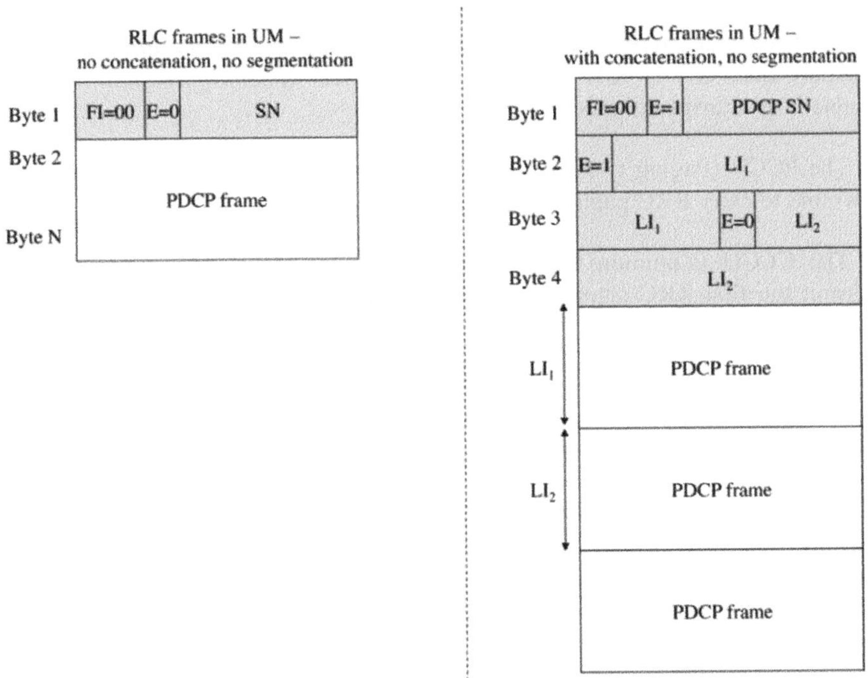

Figure 2.3. *Structure of the RLC frame – UM*

FI (Framing Information): this field is coded on 2 bits. It indicates whether or not segmentation is used:

– FI = 00, no segmentation;

– FI = 01, start of segmentation of the last PDCP frame;

– FI = 10, end of segmentation of the first PDCP frame;

– FI = 11, segmentation of the first and last PDCP frame.

E (Extension): this bit indicates the presence (bit value of ONE) or absence (bit value of ZERO) of an extension of the RLC header after the first field (SN or LI).

SN (Sequence Number): this field is coded on 5 or 10 bits. It contains the sequence number of the RLC frames. Numbering of the RLC frames helps to re-sequence the frames received from the MAC sub-layer.

LI (Length Indicator): this field is coded on 11 bits. It is used only during concatenation. It contains the length of the corresponding PDCP frame or the segment. The LI field is present for all frames except the last.

In AM, the structure of the RLC header depends solely on the number of PDCP frames concatenated, with the SN field having a fixed length of 10 bits. Figure 2.4 shows the RLC header for the following two cases:

– no concatenation of PDCP frames;

– segmentation of a PDCP frame and concatenation of two PDCP frames.

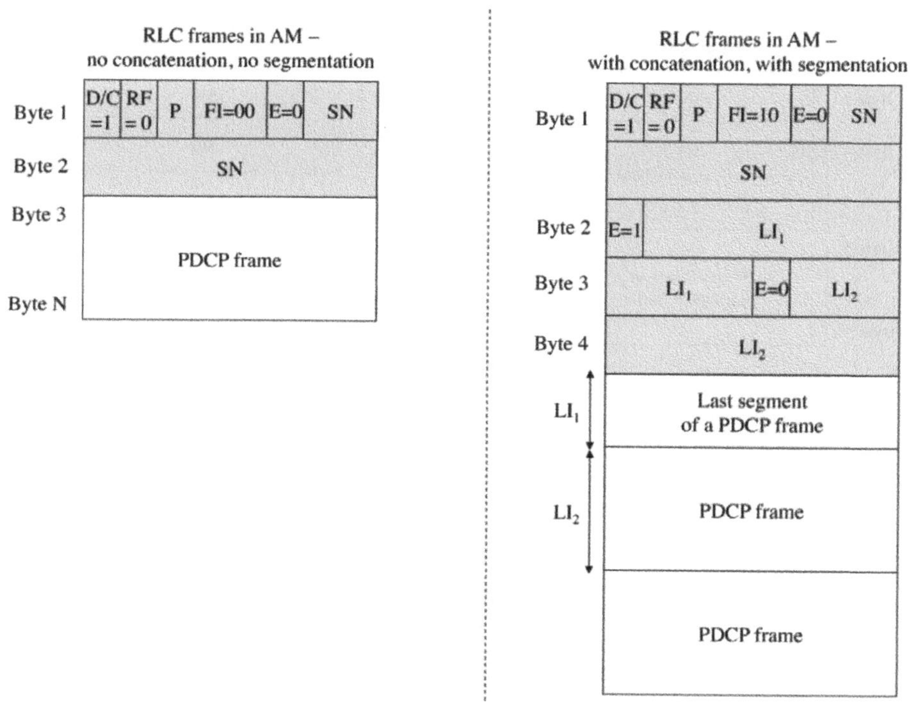

Figure 2.4. *Structure of the RLC frame – AM transmission of the initial frame*

D/C (Data/Control): this bit indicates whether the frame contains PDCP frames (bit value of ONE) or control messages (bit value of ZERO).

RF (Re-segmentation Flag): this bit indicates whether re-transmission takes place with (bit value of ONE) or without (bit value of ZERO) re-segmentation.

P (Polling): this bit indicates a request relating to an acknowledgement of the frames received.

In the case of a re-transmission of an RLC frame for which the initial resource is unavailable, the PDCP frame initially transmitted has to be able to be segmented. Figure 2.5 shows the RLC header for the following two cases:

– re-transmission of a PDCP frame in the form of a segment;

– re-transmission of three PDCP frames, with the first PDCP frame in the form of a segment.

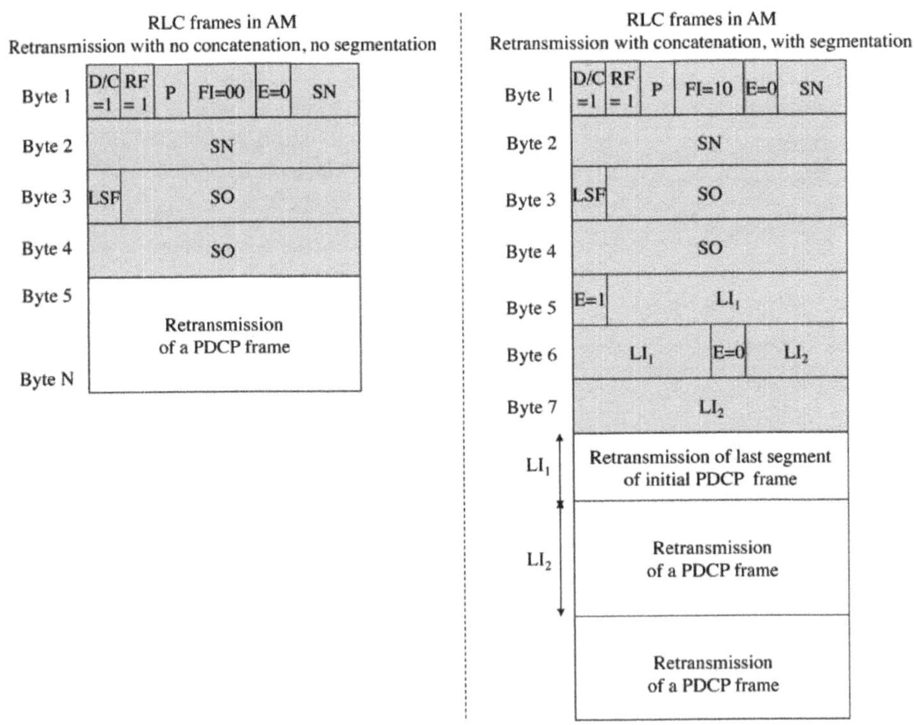

Figure 2.5. *Structure of the RLC frame – AM re-transmission of the initial frame*

LSF (Last Segment Flag): the bit indicates that the segment corresponds to the end of the initial PDCP frame.

SO (Segment Offset): this field coded on 15 bits indicates the position of the segment in the initial PDCP frame. It contains the current number of the first byte in the segment.

Figure 2.6 shows the control message for the RLC protocol. It relates to the acknowledgement of RLC frames and indication of lost frames and frame segments.

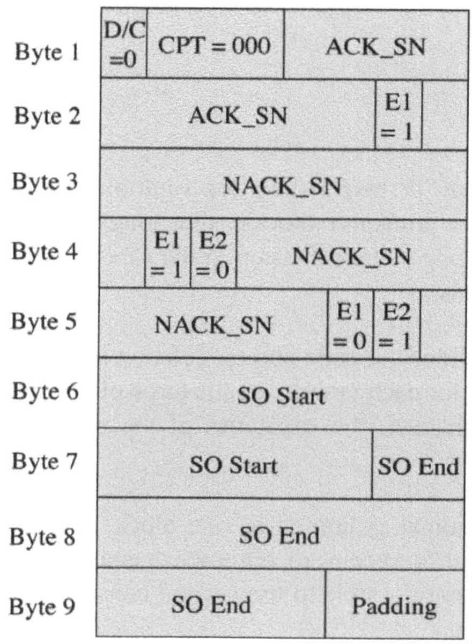

Figure 2.6. *Control message for RLC protocol*

CPT (Control PDU Type): this field is coded on 3 bits. It indicates the type of control message. The only message defined relates to the acknowledgement report for the RLC frames received.

ACK_SN (Acknowledgement SN): this field is coded on 10 bits. It indicates the sequence number of the next expected frame. All the frames transmitted with an SN lower than that number are validated, with the exception of frames whose SN appears in the field NACK_SN.

E1 (Extension 1): this bit indicates the presence (bit value of ONE) or absence (bit value of ZERO) of the fields NACK_SN, E1 and E2 thereafter.

NACK_SN (Negative Acknowledgement SN): this field is coded on 10 bits. It contains the SN of the lost frame.

E2 (Extension 2): this bit indicates the presence (bit value of ONE) or absence (bit value of ZERO) of the fields SO Start and SO End thereafter.

SO Start: this field is coded on 15 bits. It indicates the number of the first byte of the lost segment.

SO End: this field is coded on 15 bits. It indicates the number of the last byte of the lost segment.

2.2.3. *MAC protocol*

The MAC protocol is responsible for multiplexing the RLC frames from different entities in the transport blocks, allocating resources using a scheduling mechanism and managing re-transmission in the case of error by way of the HARQ (Hybrid ARQ) mechanism.

The scheduling defines the code and modulation scheme that need to be applied to the transport block, for each mobile, on the basis of the conditions of transmission on the radioelectric channel, the capacities of the mobiles and the level of QoS required.

HARQ uses a windowing limited to one block, with the sending of one data block being conditional on receipt of acknowledgement of the previous block. This procedure means it is not possible to use several consecutive TTIs, which leads to a decrease of the datarate.

The way in which this problem is solved is to establish several parallel instances of HARQ for the same user. While one instance is awaiting acknowledgement of one transmitted data block, the scheduling mechanism can assign the transmission of the next block to another instance. It should be noted that the HARQ mechanism may lead to the de-sequencing of the RLC frames received if re-transmitted.

2.2.3.1. *Transport channels*

The BCH (Broadcast Channel) supports the logic channel BCCH, containing the system information corresponding to the RRC message *MasterInformationBlock*.

The PCH (Paging Channel) transports the logic channel PCCH. It supports discontinuous transmission at pre-defined times so as to help save the mobile battery.

The DL-SCH (Downlink Shared Channel) is the main channel used for downlink. It can multiplex all of the logic channels. It supports dynamic datarate adaptation and HARQ.

The MCH (Multicast Channel) transports the multicast traffic data and their associated control data.

The RACH (Random Access Channel) does not transport a logic channel. It is used by the mobile for random access when the mobile is not connected to the eNB. It can also be used when the mobile changes cell or when it wishes to transmit but no resource is attributed, or when it has lost time synchronization. The RACH only transports a preamble to initialize the connection with the eNB.

The UL-SCH (Uplink Shared Channel) is the main channel used for uplink. It transports the logic channels for signaling (DCCH, CCCH) and traffic (DTCH). It supports dynamic datarate adaptation and HARQ.

2.2.3.2. *Protocol structure*

The MAC protocol defines a header for multiplexing RLC frames from different instances and control elements for the protocol itself. The global header for MAC therefore comprises a collection of unitary headers, with each unitary header relating to an RLC frame or a control element. The order of the RLC frames or the control elements corresponds to that of the unitary headers. The headers relating to the control elements come before the headers relating to the RLC frames.

Figure 2.7 illustrates the general structure of a MAC frame. Each unitary header has a size of 2 or 3 bytes depending on the field L (Length).

E (Extension): this bit indicates whether the next unitary header is present (bit value of ONE) or not (bit value of ZERO).

LCID (Logic Channel Identifier): this field is coded on 5 bits. It gives the identifier of the instance of the logic channel or the type of control element (Table 2.1).

F (Format): this bit indicates the format of the field L (Length), coded either on 7 bits (bit F value of ONE) or 15 bits (bit F value of ZERO).

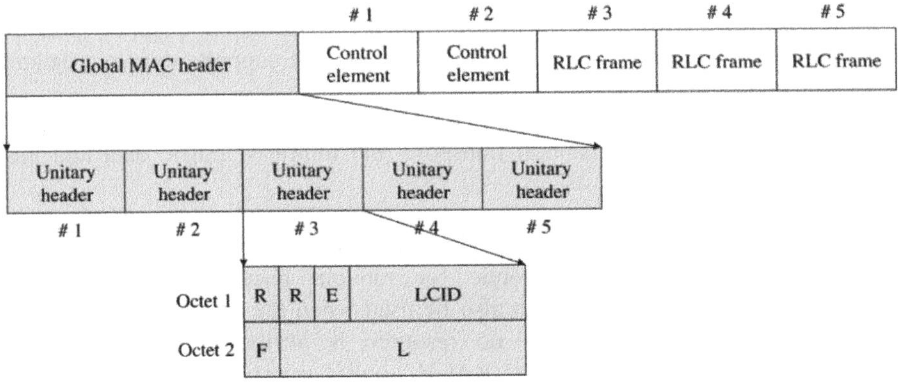

Figure 2.7. *Structure of MAC frame*

Index	Value
Uplink	
Logic channel CCCH	00000
Logic channel ID	00001 to 01010
Reserved	01011 to 11011
ID of the mobile for contention resolution	11100
Timing Advance (TA)	11101
Control of idle mode DRX (Discontinuous Reception)	11110
Padding	11111
Downlink	
Logic channel CCCH	00000
Logic channel ID	00001 to 01010
Reserved	01011 to 11001
Power Headroom	11010
C-RNTI (cell radio network temporary identifier)	11011
Buffer status (truncated report)	11100
Buffer status (short report)	11101
Buffer status (long report)	11110
Padding	11111

Table 2.1. *Values for the field LCID*

The MAC frame RAR (Random Access Response) is a special frame used by the eNB in response to receiving an attempt to connect by the mobile. As before, the global header contains a collection of unitary headers. The first header may be a special header giving an indication of the cell's load.

RAR messages contain the Timing Advance (TA) value, indications about the resource attributed to the mobile for transmission for uplink and the C-RNTI attributed to the RRC signaling.

Figure 2.8 shows the general structure of a MAC frame RAR with indication of cell load.

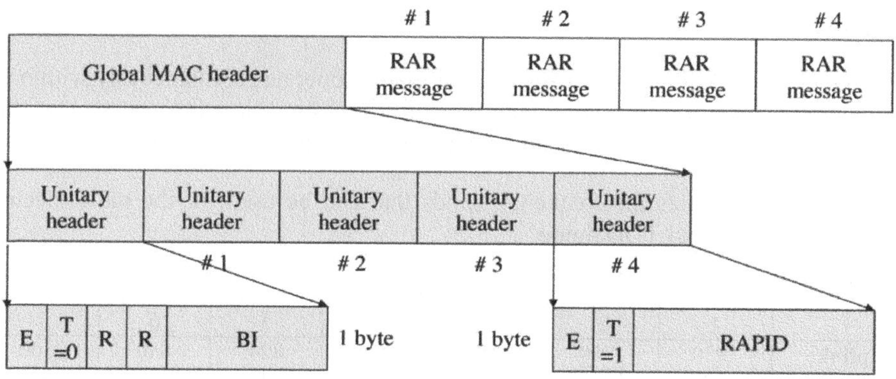

Figure 2.8. *Structure of the MAC frame RAR*

E (Extension): this bit indicates whether the next unitary header is present (bit value of ONE) or not (bit value of ZERO).

T (Type): this bit gives an indication of the structure of the unitary header (T=0 for BI, T=1 for RAPID).

RAPID (Random Access Preamble Identifier): this field is coded on 6 bits. It contains the preamble identifier used by the mobile to connect to the eNB.

BI (Backoff Indicator): this field is coded on 4 bits. It gives an indication of the cell load.

2.3. Physical layer

Two major functions dictate the structure of the physical layer:

– coding of the transport channel to facilitate error correction and residual error detection. The physical channel is the data structure resulting from this treatment;

– time and frequency multiplexing of the different physical channels. The physical channel transporting the traffic data (IP packet) and signaling data (RRC messages) from the mobile is randomly assigned a resource comprising an amount of time (1 ms) and frequency ($n \times 180$ kHz).

2.3.1. *Frequency range*

The transmission in both directions uses two coupled bandwidths in FDD (Frequency Division Duplex) mode or a single bandwidth in TDD (Time Division Duplex) mode.

For FDD mode, the two transmission directions operate simultaneously within an assigned bandwidth. For TDD mode, both transmission directions operate on the same bandwidth, with each direction being allocated a portion of the time.

Table 2.2 indicates the frequency bands that can be used for the radio electric interface, in FDD and TDD mode.

FDD Mode							
	1	2	3	4	5	6	7
Uplink (MHz)	1920 1980	1850 1910	1710 1785	1710 755	824 849	830 840	2500 2570
Downlink (MHz)	2110 2170	1930 1990	1805 1880	2110 2155	869 894	875 885	2620 2690
FDD Mode							
	8	9	10	11	12	13	14
Uplink (MHz)	880 915	1750 1785	1710 1770	1428 1463	698 716	777 787	788 798
Downlink (MHz)	925 960	1845 1880	2110 2170	1476 1501	728 746	746 756	758 768
TDD Mode							
	33	34	35	36	37	38	39
Uplink / downlink (MHz)	1900 1920	2010 2025	1950 1910	1930 1990	1910 1930	2570 2620	1880 1920

Table 2.2. *Frequency bands in FDD and TDD mode*

The frequency bands are distributed between the different continents:

– Europe and Asia (including Japan and China): bands 1, 3, 7 and 8 in FDD and 33 (except Japan), 34, 38 (except Asia) and 40 in TDD;

– America: bands 2, 4, 5, 10, 12, 13 and 14 in FDD;

– Japan: bands 6, 9 and 11 in FDD;

– China: band 38 in TDD.

The width of the frequency band, or bandwidth, is made up of sub-carriers with a 15 kHz or 7.5 kHz step. The number of sub-carriers is dependent on the bandwidth of the radioelectric channel (Table 2.3).

Bandwidth of the radio channel	1.4 Mhz	3 Mhz	5 Mhz	10 Mhz	15 Mhz	20 Mhz
Bandwidth used	1.08 Mhz	2.7 Mhz	4.5 Mhz	9 Mhz	13.5 Mhz	18 Mhz
Number of sub-carriers $\Delta F = 15$ kHz	72	180	300	600	900	1200

Table 2.3. *Characteristics of bandwidth*

The same frequency band can be used in every cell covered by the eNB. To combat interference, the following device is put in place (Figure 2.9):

– all of the spectrum is used in the central area of the cell, where interference created by the neighboring cells is slight;

– only part of the spectrum is used on the periphery of the cell, which helps eliminate interference from the neighboring cells. The mechanism ICIC (Inter-Cell Interference Coordination) dynamically distributes the spectrum between the cells depending on the traffic.

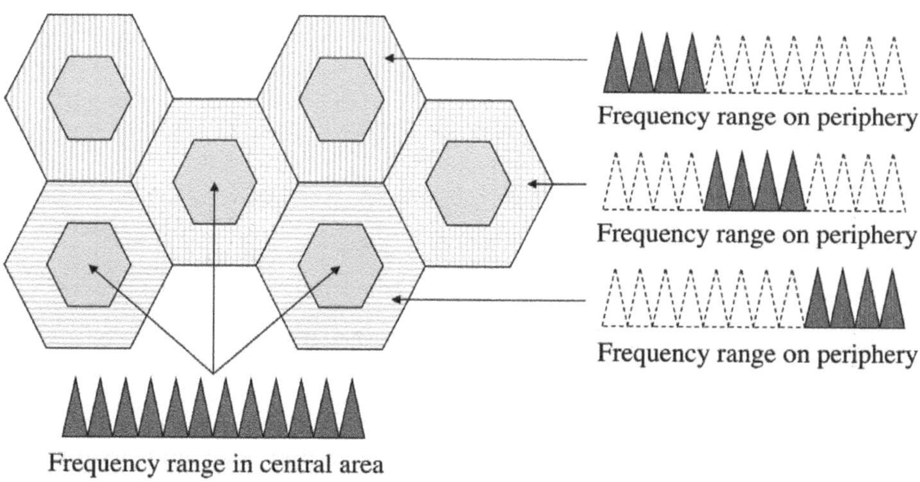

Figure 2.9. *Frequency range used in cells*

2.3.2. *Spatial multiplexing*

The system of antennas is characterized by four possible modes of transmission on the radioelectric channel. The terms "input" and "output" apply to data entering or exiting the radioelectric channel.

SISO (Single Input Single Output) mode is the basic signal propagation mode for which a transmitting antenna and a receiving antenna are used (Figure 2.10).

SIMO (Single Input Multiple Output) mode is characterized by the use of one transmitter and multiple receivers (Figure 2.10). SIMO mode is often spoken of in terms of diversity of reception. The transmitted datarate is identical to that in SISO mode. However, the combination of the different received signals helps the receiver combat the problem of fading of the radioelectric signal.

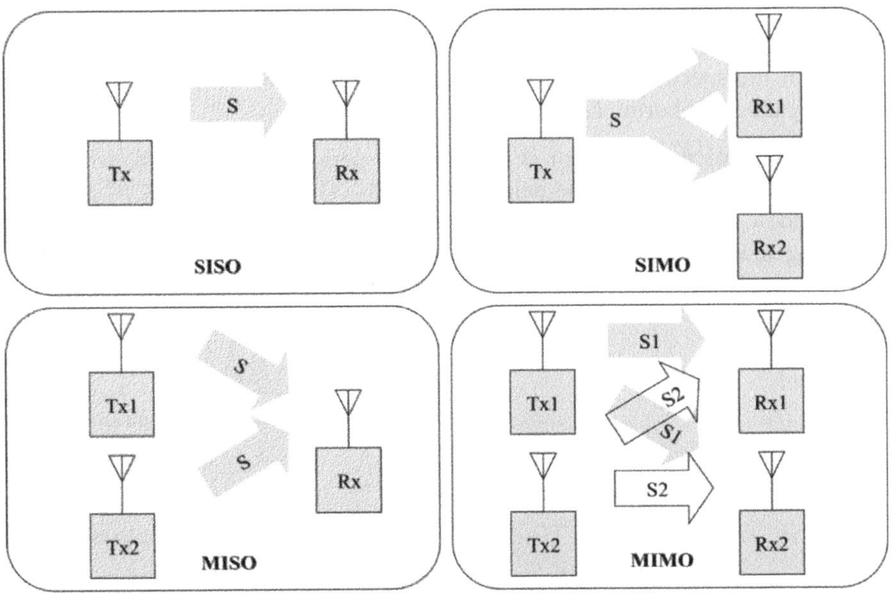

Figure 2.10. *The different antenna systems*

MISO (Multiple Input Single Output) mode uses several transmitters and a single receiver (Figure 2.10). The same signal is transmitted from each of the transmitting antennas. MISO mode is often spoken of in terms of diversity of transmission. Similarly to SIMO mode, MISO mode helps the receiver deal with the problem of fading of the radioelectric signal.

MISO mode is also used to logically modify the beam of the antenna and thereby improve the signal-to-noise ratio (SNR) by controlling the phases of the different signals emitted (a technique known as beamforming).

MIMO (Multiple Input Multiple Output) mode uses several antennas both for transmission and reception (Figure 2.10). It improves the datarate by facilitating the transmission of different signals on the same frequency at the same time.

Seven modes of transmission are defined:

1 – SIMO mode;

2 – MISO mode, for downlink. This mode uses SFBC (Space Frequency Block Code) if two transmitters are used. If four transmitting antennas are used, the SFBC mechanism is supplemented by an additional diversity mechanism – FSTD (Frequency Shift Transmit Diversity);

3 – SU-MIMO (Single-User MIMO) mode, for downlink, without precoding. This mode requires the mobile to be equipped with two receivers (2 × 2 MIMO) or four receivers (4 × 4 MIMO);

4 – SU-MIMO mode, for downlink, with precoding. This precoding enables us to define the phase of the transmitted signal on the basis of the values from a dictionary;

5 – MU-MIMO (Multi-User MIMO) mode, for uplink. Each mobile has only one transmitter, so no modification needs to be made to the mobile. The mobile's datarate therefore remains unchanged. The different transmission antennas are those of different mobiles. The eNB, on the other hand, has two receivers (2 × 2 MIMO) or four (4 × 4 MIMO). This mode increases the cell's capacity;

6 – SU-MIMO mode, with precoding limited to a single physical layer;

7 – MISO mode, for downlink, with an additional antenna on the eNB and precoding unique to the user for downlink. The mobile controls the phasing of the two signals received from the eNB, which causes a modification in the shape of the beam from the antenna (beamforming) and increases the range.

2.3.3. *Time multiplexing*

Two structures for frames are defined, depending on whether FDD mode or TDD mode is used. For uplink, the signals transmitted by the different mobiles need to be temporally aligned in a sub-frame, when the eNB receives them. Therefore, the mobiles need to be time-synchronized by the eNB, which informs them of the timing advance (TA) to be applied.

The type-1 structure defined for FDD mode has a duration of 10 ms and contains 10 sub-frames (Figure 2.11). Each sub-frame comprises two time-slots. Each time-slot contains 3, 6 or 7 OFDM (Orthogonal Frequency-Division Multiplexing) symbols (Figure 2.14). A symbol contains a number of bits corresponding to the type of modulation used:

– 2 bits with QPSK modulation (Quadrature Phase-Shift Keying), whose constellation contains 4 different symbols;

– 4 bits in the case of 16-QAM (Quadrature Amplitude Modulation), whose constellation contains 16 different symbols;

– 6 bits in the case of 64-QAM, whose constellation contains 64 different symbols.

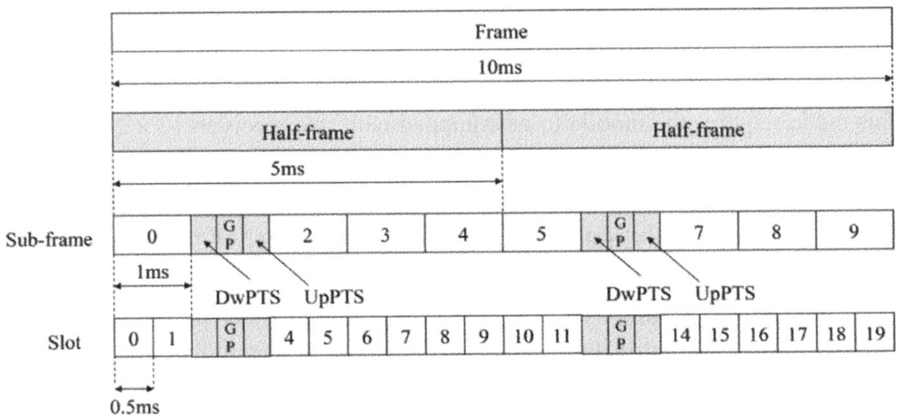

Figure 2.11. *Structure of the frame in FDD mode*

The type-2 structure defined for TDD mode also has a duration of 10 ms and contains two half-frames of 5 ms each (Figure 2.12). Each half-frame includes 5 time slots, the second of which contains three particular fields:

– a field for the downlink pilot DwPTS (Downlink Pilot Time Slot). This field may contain data;

– a field for the uplink pilot UpPTS (Uplink Pilot Time Slot). This field may contain data or a preamble;

– a GP (Gap Period) between the previous two fields. This time interval helps to compensate for a time delay between different mobiles and thereby to prevent overlap between the two transmission directions.

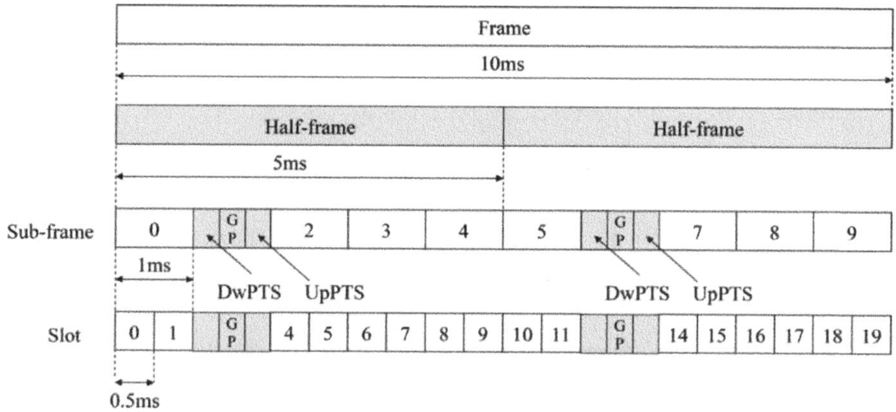

Figure 2.12. *Frame structure in TDD mode*

The sub-frames are attributed to data for uplink and downlink in various configurations (Table 2.4):

– sub-frames 0 and 5 are always allocated to downlink traffic;

– sub-frame 1 is always allocated to the three special fields;

– sub-frame 2 is always allocated to uplink traffic;

– sub-frame 6 is allocated to the three special fields for a periodicity of 5 ms;

– sub-frames 3, 4, 7, 8 and 9 are allocated to either downlink or uplink traffic depending on the configuration chosen.

Configuration	Periodicity	Sub-frame number									
		0	1	2	3	4	5	6	7	8	9
0	5 ms	D	S	U	U	U	D	S	U	U	U
1	5 ms	D	S	U	U	D	D	S	U	U	D
2	5 ms	D	S	U	D	D	D	S	U	D	D
3	10 ms	D	S	U	U	U	D	D	D	D	D
4	10 ms	D	S	U	U	D	D	D	D	D	D
5	10 ms	D	S	U	D	D	D	D	D	D	D
6	5 ms	D	S	U	U	U	D	S	U	U	D

D (Downlink)/U (Uplink)/S (Special)

Table 2.4. *TDD frame configuration*

Figure 2.13 shows the structure of the TDD frame for configuration 1.

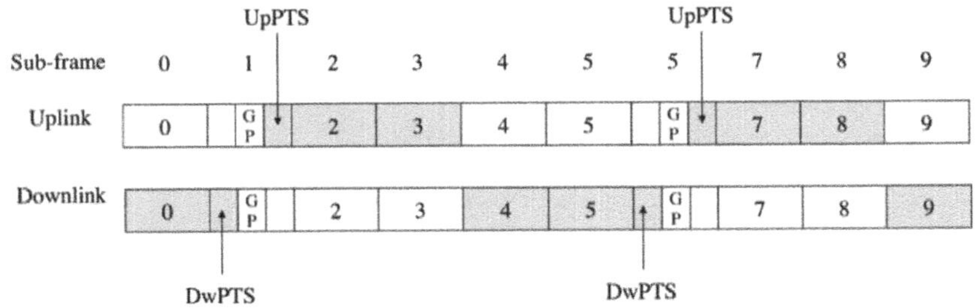

Shading represents the sub-frames allocated to each transmission direction

Figure 2.13. *TDD frame structure – configuration 1*

By including a guard time-period between the symbols, it is possible to eliminate inter-symbol interference. The guard time for a slot is located at the beginning of the slot. It contains the copy of the end of the symbol in order to prevent the dynamics from being too great for the amplifiers. This copy is called the cyclic prefix (Figure 2.14).

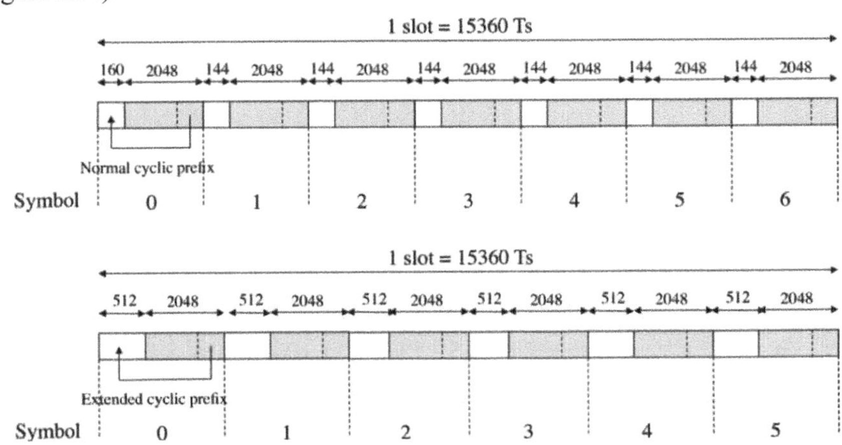

Figure 2.14. *Time slot structure*

The normal cyclic prefix is used if the separation between the different reflected signals is slight, which is the case with small-diameter cells (see Figure 2.14 and Table 2.5).

The extended cyclic prefix is used essentially when the extent of the delay between the different reflected signals is great, which is the case with large-diameter cells (Figure 2.14 and Table 2.5).

Downlink		Cyclic prefix length
Normal cyclic prefix	$\Delta f = 15$ kHz	160 × Ts where $l = 0$ 144 × Ts where $l = 1, 2, ..., 6$
Extended cyclic prefix	$\Delta f = 15$ kHz	512 × Ts where $l = 0, 1, ..., 5$
	$\Delta f = 7.5$ kHz	1024 × Ts µs where $l = 0, 1, 2$
Uplink		**Cyclic prefix length**
Normal cyclic prefix		160 × Ts where $l = 0$ 144 × Ts where $l = 1, 2, ..., 6$
Extended cyclic prefix		512 × Ts where $l = 0, 1, ..., 5$

Δf: step between sub-carriers
l: number of symbol in slot

Table 2.5. *Cyclic prefix length*

A Resource Element (RE) is the smallest unit that can be attributed to a physical signal (Figure 2.15). It corresponds to one OFDM symbol on the temporal axis and to one sub-carrier on the frequency axis.

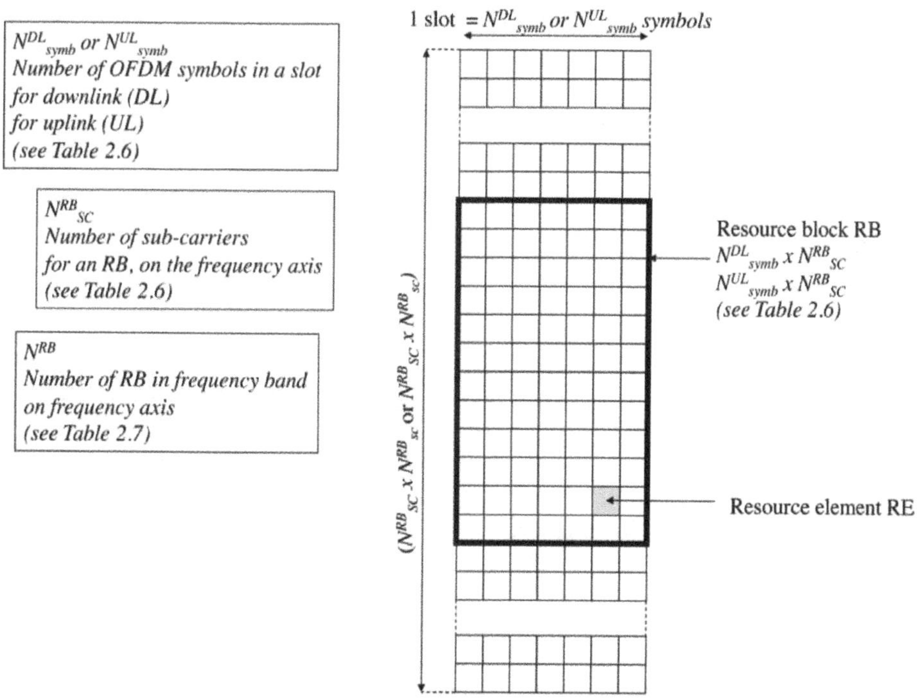

Figure 2.15. *Structure of resource block*

The Resource Block (RB) is the unit of resources attributed to a mobile (Figure 2.15). It represents 0.5 ms (1 slot) on the temporal axis and 180 kHz on the frequency axis. On the temporal axis, two RBs are attributed to a mobile. The number of sub-carriers and the number of symbols per RB depend on the space between the sub-carriers and the size of the cyclic prefix (Table 2.6).

For a normal normal cyclic prefix, an RB contains 84 REs (7 OFDM symbols × 12 sub-carriers) (Figure 2.15 and Table 2.6). For an extended cyclic prefix, an RB contains 72 REs (6 OFDM symbols × 12 sub-carriers or 3 OFDM symbols × 24 sub-carriers) (Table 2.6).

Downlink		N^{RB}_{sc}	N^{DL}_{symbol}
Normal cyclic prefix	Δf = 15 kHz	12	7
Extended cyclic prefix	Δf = 15 kHz		6
	Δf = 7.5 kHz	24	3
Uplink		N^{RB}_{sc}	N^{UL}_{symbol}
Normal cyclic prefix		12	7
Extended cyclic prefix		12	6

N^{RB}_{sc} – number of sub-carriers per RB – frequency axis
N^{DL}_{symbol} – number of symbols per slot – downlink
N^{UL}_{symbol} – number of symbols per slot – uplink

Table 2.6. *Parameters of the resource block*

The number of RBs on the frequency axis is a function of the bandwidth of the radioelectric channel (Table 2.7).

Bandwidth (MHz)	1.4	3	5	10	15	20
Bandwidth used (MHz)	1.08	2.7	4.5	9	13.5	18
Number of RBs	6	15	25	50	75	100

Table 2.7. *Number of resource blocks on frequency axis*

2.3.4. *Physical signals and channels*

The radioelectric interface includes physical signals and physical channels. The physical signals are used as for the synchronization of the receiver, the determination of the Physical Cell Identifier (PCI) and the assessment of the radioelectric channel. The physical channels serve to transport the traffic data and signaling data. The physical layer also has its own physical control channels.

2.3.4.1. *Downlink*

The downlink physical signals and channels are shown in Table 2.8.

Physical signals		
PSS	Primary Synchronization Signal	Synchronization of the half-frame and physical cell identification
SSS	Secondary Synchronization Signal	Synchronization of the frame and physical cell identification, FDD or TDD operation, cyclic prefix length
RS	Reference Signal	Frequency synchronization and assessment of the radioelectric channel
Physical channels		
PBCH	Physical Broadcast Channel	System information about the cell
PDCCH	Physical Downlink Control Channel	Signaling regarding the allocation of the data received during downlink and resource allocation for the data to be uplinked
PCFICH	Physical Control Format Indicator Channel	Number of OFDM symbols attributed to PDCCH
PHICH	Physical HARQ Indicator Channel	Acknowledgement of data received from the mobile over the PUSCH channel
PDSCH	Physical Downlink Shared Channel	Support of the transport channel for unicast traffic and signaling
PMCH	Physical Multicast Channel	Support of the traffic transport multicast channel MCH

Table 2.8. *Physical signals and channels – downlink*

2.3.4.1.1. PSS

A cell has a physical identifier which can assume 504 values, distributed in 168 groups of 3 values. The Primary Synchronization Signal (PSS) is generated using a Zadoff–Chu sequence corresponding to the value of one of the three groups.

In FDD mode, the PSS is mapped in the last symbol of slots 0 and 10. It occupies 62 resource elements situated on both sides of the central frequency (Figure 2.16).

The sequence is the same for the two slots, which enables us to recover synchronization of the half-frame.

The position of the physical PSS is independent of the bandwidth of the radioelectric channel.

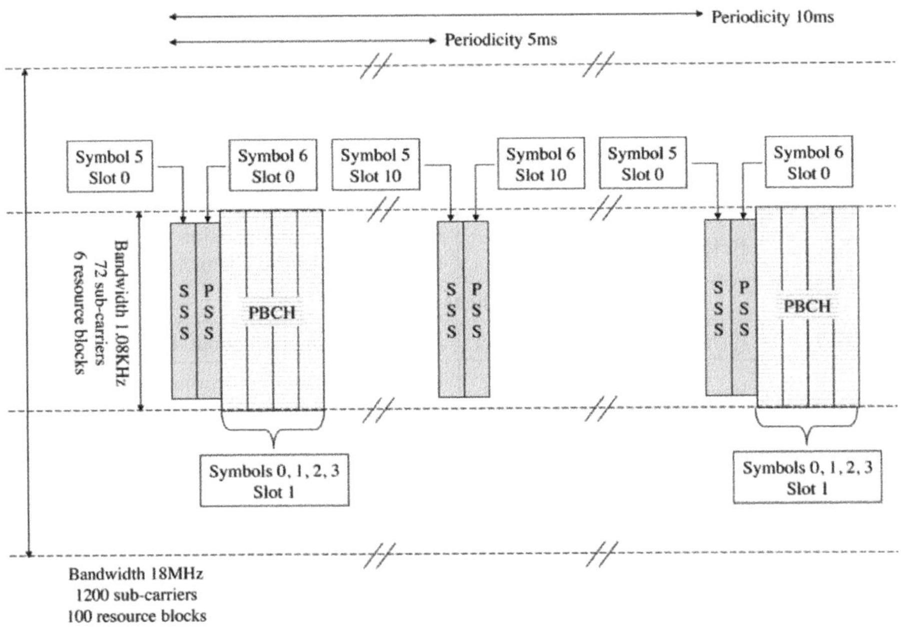

Figure 2.16. *Mapping of the PSS, SSS and PBCH channels*

2.3.4.1.2. SSS

The physical Secondary Synchronization Signal (SSS) corresponds to two sequences of 31 interleaved bits, which indicates the value of the group (1 value out of 168).

In FDD mode, the SSS is mapped in the penultimate symbol of slots 0 and 10. It occupies 62 resource elements situated on both sides of the central frequency (Figure 2.16).

The combination of the two sequences of 31 bits is different for each slot, which enables us to recover synchronization of the frame.

The position of the SSS is independent of the bandwidth of the radioelectric channel.

2.3.4.1.3. RS

There are two types of Reference Signal (RS):

– the RS dedicated to a cell, transmitted over the four antennas;

– the RS dedicated to a mobile, transmitted over an additional antenna.

For the first two antennas, the RS dedicated to a cell is mapped in four REs in each RB. For the next two antennas, the RS is mapped in two REs. The REs of the RS for one antenna must not be re-used for the other antennas (Figure 2.17).

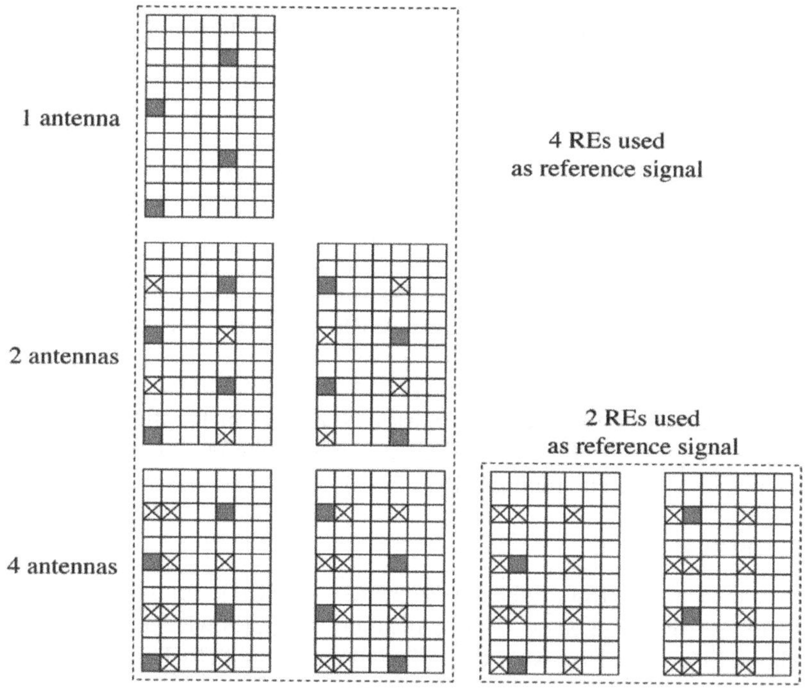

Figure 2.17. *Mapping of the RS dedicated to the cell*

The position of the RS depends on the PCI.

A frequency shift is applied to the RS so as to avoid interference between neighboring cells.

The RS dedicated to a cell is generated using a pseudo-random sequence of 31 bits (Gold code), the initial value of which depends on the number of the time slot and the PCI.

The RS dedicated to a mobile is transmitted only in the RBs attributed to the mobile. It is mapped in six REs in each RB. It is generated on the basis of a pseudo-random sequence of 31 bits (Gold code), the initial value of which depends on the number of the time slot, the PCI and the C-RNTI.

2.3.4.1.4. PBCH

The Physical Broadcast Channel (PBCH) occupies the first four symbols of time slot 1. The information is divided over four consecutive frames. The PBCH is mapped in 72 REs situated on both sides of the central frequency (Figure 2.16).

The position of the physical channel PBCH is independent of the bandwidth of the radioelectric channel.

2.3.4.1.5. PCFICH

The Physical Control Format Indicator Channel (PCFICH) transports the two bits of the CFI (Control Format Indicator), coding the number of symbols attributed to the physical control channels.

The first symbol of each sub-frame is decomposed into two REGs (Resource Element Groups). Each group contains 4 REs, ignoring the two REs that are immobilized for the RS.

The PCFICH is transmitted in 16 REs (4 REGs) in the first symbol in the sub-frame. The position of the four REGs depends on the PCI, to avoid interference between neighboring cells (Figure 2.18).

2.3.4.1.6. PHICH

The Physical HARQ Indicator Channel (PHICH) transports the bit of information HARQ Indicator (HI), indicating an acknowledgement ACK or a non-acknowledgement NACK of the uplinked data. The PHICH group multiplexes eight HIs.

The PHICH group is transmitted in 12 REs (3 REGs) located in the first symbol of the sub-frame. The position of the 3 REGs depends on the PCI (Figure 2.18).

2.3.4.1.7. PDCCH

The Physical Downlink Control Channel (PDCCH) is transmitted in the first symbols of each sub-frame. The number of symbols used is indicated by the PCFICH channel.

The PDCCH is divided into the REGs available in the first symbol and the REGs of the subsequent symbols (Figure 2.18).

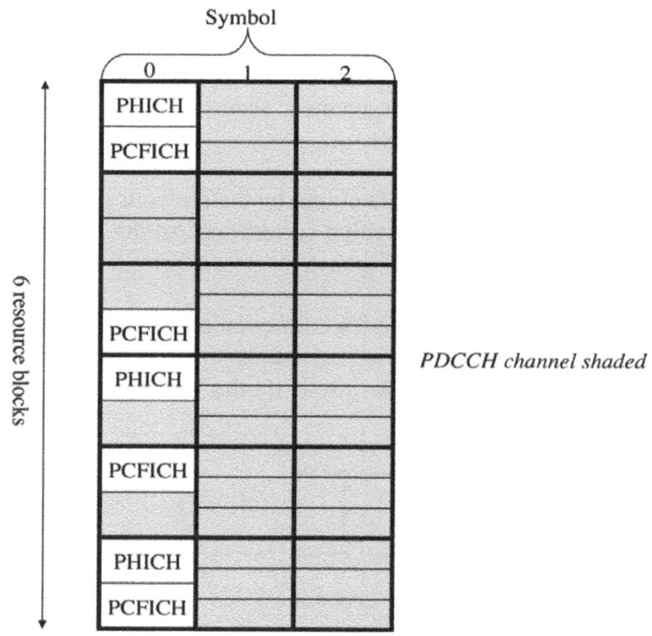

Figure 2.18. *Mapping of the control channels PCFICH, PHICH and PDCCH*

The 12 REs of the second (only in the case of use of one or two antennas), third and fourth symbol are divided into three REGs, with each group having a capacity of four resource elements. If four antennas are used, each RB of the second symbol is divided into two REGs, in order to take account of the presence of reference signals.

The physical channel PDCCH transports different reports of DCI (Downlink Control Information), distributed in four types of format:

– type 0 contains the information relating to resource attribution for uplink;

– types 1a, 1b, 1c and 1d contain the information relating to resource attribution for downlink, where only one transport block is used;

– types 2 and 2b contain the information relating to resource attribution for downlink, where two transport blocks are used;

– types 3 and 3a contain the information relating to control of the power of the physical uplink channels, PUSCH and PUCCH.

2.3.4.1.8. PDSCH

One or two transport blocks are processed, with each block constituting a codeword. Figure 2.19 shows the functions applied to each codeword.

A 24-bit Cyclic Redundancy Code (CRC) is applied to the codeword to detect any residual errors in reception. The receiver sends a HARQ ACK if the data are correct and a HARQ NACK otherwise.

Segmentation is applied to the data output by the previous phase if the size of the data is greater than 6144 bits. This value corresponds to the maximum size of the turbo-code.

A turbo-code is used to perform error correction on reception, with a coding rate of 1/3. The datarate adaptation function punctures bits in order to establish multiple redundant versions which will be used if the HARQ function requires a re-transmission.

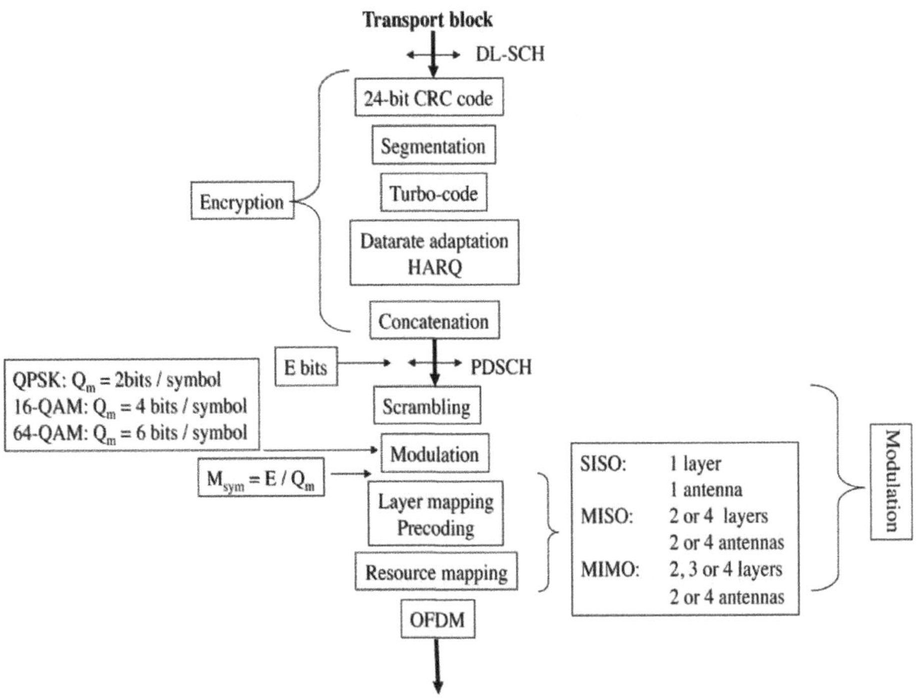

Figure 2.19. *The physical channel PDSCH*

If segmentation has been carried out, the different segments are concatenated before being scrambled by a sequence specific to the cell. The data are scrambled so as to reduce the impact of interference. The scrambling sequence depends on the PCI, the number of the time slot, the value of the RNTI and the number of the codeword.

Depending on the type of modulation (QPSK, 16-QAM or 64-QAM), each symbol contains 2, 4 or 6 bits. The symbols are mapped on several layers, to which precoding is applied. The symbols are then mapped on the antennas.

Downlink uses the mechanism OFDM, comprising several orthogonal sub-carriers, spaced 15 kHz apart.

The IFFT (Inverse Fast Fourier Transform) enables us to switch from a frequential representation (symbols) to a temporal representation.

The Physical Downlink Shared Channel (PDSCH) occupies the REs that are not used by the physical signals and the physical control channels.

Table 2.9 shows the peak datarates for the PDSCH on the basis of the following hypotheses:

– 64 QAM modulation is used;

– a symbol is attributed to the PDCCH control channels;

– the coding rate is 1;

– the computation does not take account of the physical synchronization signals (PSS, SSS) or of the physical channel PBCH.

Frequency bandwidth	**Number of antennas**		
	1	**2**	**4**
1.4 Mhz	5.4 Mbit/s	10.4 Mbit/s	19.6 Mbit/s
3 Mhz	13.5 Mbit/s	25.9 Mbit/s	50 Mbit/s
5 Mhz	22.5 Mbit/s	43.2 Mbit/s	81.6 Mbit/s
10 Mhz	45 Mbit/s	86.4 Mbit/s	163.2 Mbit/s
15 Mhz	67.5 Mbit/s	129.6 Mbit/s	244.8 Mbit/s
20 Mhz	90 Mbit/s	172.8 Mbit/s	326.4 Mbit/s

Table 2.9. *Datarates of the physical channel PDSCH*

2.3.4.2. *Uplink*

The physical signals and channels for uplink are shown in Table 2.10.

Physical signals		
DRS	Demodulation Reference Signal	Frequency synchronization
SRS	Sounding Reference Signal	Assessment of the radioelectric channel
Physical channels		
PRACH	Physical Random Access Channel	Random access to the radioelectric resource
PUCCH	Physical Uplink Control Channel	Signaling relating to acknowledgement of data received during downlink, the quality of the signal received and the scheduling request
PUSCH	Physical Uplink Shared Channel	Support of the unicast traffic transport channel

Table 2.10. *Physical signals and channels – uplink*

2.3.4.2.1. Physical reference signals

The Sounding Reference Signal (SRS) is generated using a Zadoff-Chu sequence, whose characteristic is that it optimizes the value of auto- and inter-correlation. Multiple SRSs coming simultaneously from different mobiles are multiplexed by these sequences.

The SRS is transmitted in the last symbol of the sub-frame and uses the whole of the spectrum, with a periodicity ranging from 2 ms (every two sub-frames) to 160 ms (every 16 frames).

The position of the Demodulation Reference Signal (DRS) attached to the physical channel PUCCH depends on the type of channel PUCCH.

The DRS associated with the physical channel PUSCH occupies the fourth symbol, if a normal cyclic prefix is used.

2.3.4.2.2. PUCCH

The Physical Uplink Control Channel (PUCCH) is used to transport UCI (Uplink Control Information) only when no resource is attributed to the mobile for the PUSCH.

Otherwise, the PUSCH is used to transport UCI.

The type of format depends on the type of information borne (Table 2.11).

Format of PUCCH	Modulation	Number of bits per sub-frame	Usage
1	None	None	Scheduling request
1a	BPSK	1	ACK / NACK
1b	QPSK	2	ACK / NACK
2	QPSK	20	CQI
2a	QPSK+BPSK	21	CQI + ACK / NACK
2b	QPSK+BPSK	22	CQI + ACK / NACK

Table 2.11. *Formats of the PUCCH*

The PUCCH is transmitted at both extremities of the spectrum.

With the 1x format, the PUCCH is transmitted in four symbols of an RB (symbols 0, 1, 5 and 6), with the three others being attributed to the DRS associated with the physical channel (Figure 2.20).

With the 2x format, the PUCCH is transmitted in five symbols of an RB (symbols 0, 2, 3, 4, 6), with the two others being attributed to the DRS associated with the physical channel (Figure 2.20).

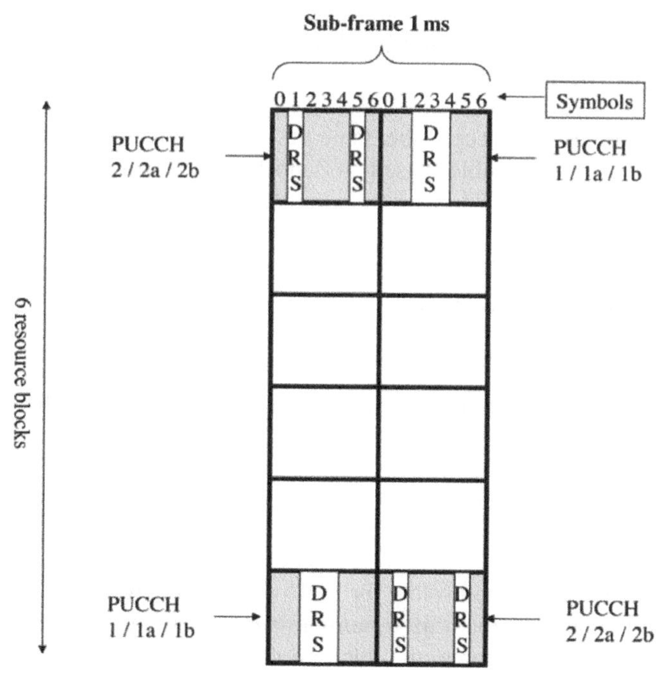

Figure 2.20. *Mapping of the PUCCH control channel*

Multiple mobiles can use the same RB. Frequency multiplexing of the control channels of different mobiles means that 12 PUCCH physical channels can be accommodated into RBs. In the case of 1x-type channels, multiplexing by the code increases the capacity of the control channel:

– 12 type-2x multiplexed channels;

– 36 type-1x multiplexed channels.

The UCI signals contain the following parameters:

– CQI (Channel Quality Indicator). This parameter gives an indication of the level of quality of the PDSCH signal received. It is used by the eNB to determine the type of modulation and code rate to apply to the PDSCH channel;

– RI (Rank Indicator). This parameter indicates the number of antennas to be used for downlink;

– PMI (Precoding Matrix Indicator). This parameter indicates the precoding values to be applied to the PDSCH channel.

2.3.4.2.3. PRACH

A resource of the Physical Random Access Channel (PRACH) occupies six RBs. The position of the PRACH is indicated by the SIB (System Information Block). The PRACH may be absent in a sub-frame or in a frame.

Multiple mobiles can access the same resource using different preambles. In each cell, there are 64 preambles based on Zadoff-Chu sequences.

The transmission of the preamble, the cyclic prefix and the guard time takes place over 1 ms, or over two time-slots, with the following distribution:

– 0.1 ms for the cyclic prefix;

– 0.8 ms for the preamble;

– 0.1 ms for the safety time, which makes it possible to have cells with a 15 km radius.

Three other structures enable us to increase the guard time when the size of the cell is large.

A fifth structure is defined solely for TDD mode. The preamble is transmitted in the field UpPTS rather than in a normal sub-frame. As the UpPTS field is transmitted only in a sub-frame, the duration of the safety time is less than that of the previous configurations. Therefore, this structure must be reserved for small cells.

2.3.4.2.4. PUSCH

The Physical Uplink Shared Channel (PUSCH) occupies the remaining REs. The PUSCH channel carries the transport blocks and possibly the control information CQI, PMI and RI (Figure 2.21).

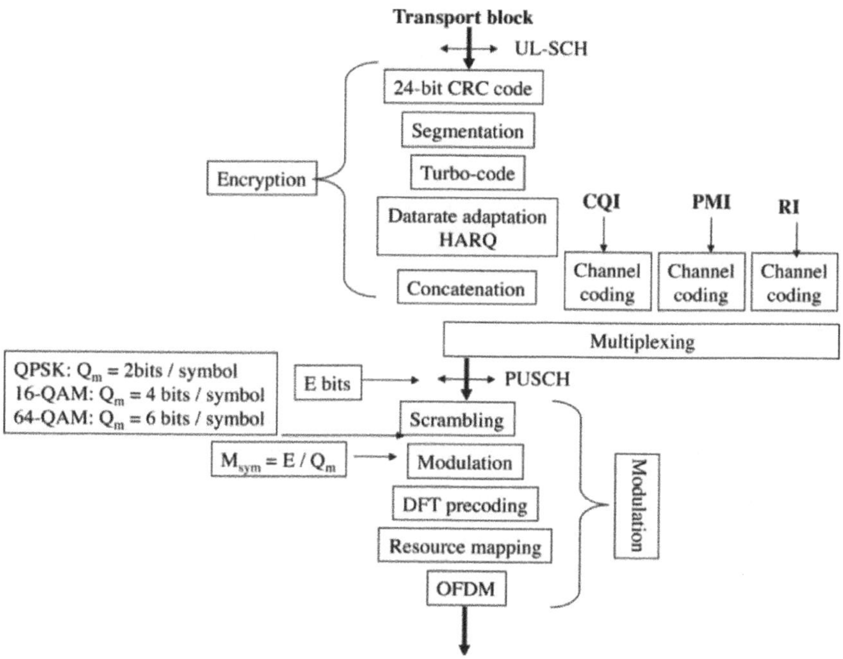

Figure 2.21. *The physical channel PUSCH*

The treatment carried out is similar to that with the PDSCH. The main difference lies in the DFT (Discrete Fourier Transform) precoding. The aim of using DFT precoding is to decrease the peak amplitude of the signal in order to optimize the consumption of the mobile.

Table 2.12 indicates the peak datarates of the PUSCH, calculated using the following hypotheses:

– 64 QAM modulation is used;

– the coding rate is 1;

– a single RB is attributed to the PUCCH for the frequency bandwidth of 1.4 Mhz, and two RBs for the other frequency bands;

– the computation does not take account of the SRS or of the PRACH.

Frequency bandwidth	1.4 Mhz	3 Mhz	5 Mhz	10 Mhz	15 Mhz	20 Mhz
Datarate (Mbit/s)	4.3	10.4	17.3	41.5	62.2	82.9

Table 2.12. *Datarates of the physical channel PUSCH*

2.4. Procedures

2.4.1. *Cell searching*

When it is switched on, the mobile acquires frequential and temporal synchronization, as well as the PCI, when it has found the physical signals PSS and SSS. The position of the physical signals in relation to the central frequency of the radioelectric channel is independent of the frequency bandwidth.

The physical signal PSS is used to retrieve synchronization of a half-frame to 5 ms and the number in the group (1 value out of 3) of the PCI.

The physical signal SSS is used to retrieve synchronization of the frame, the group number of the PCI (1 value out of 168), to find the type of cyclic prefix (normal or extended) and the type of duplex (FDD or TDD).

2.4.2. *System information*

Based on the PCI, the mobile determines the position of the reference signal RS. It is then able to process the physical channel PBCH to retrieve the MIB (Master Information Block) data which indicate the bandwidth of the radioelectric channel for downlink, the characteristic of the channel PHICH and the system frame number SFN.

On the basis of the PCI, the mobile can analyze the physical channel PCFICH to find the size of the physical channel PDCCH. Analysis of the PDCCH yields up the code SI-RNTI indicating the presence of SIB (System Information Block) data in the physical channel PDSCH. More specifically, the SIB2 information gives the characteristics of the physical channel PRACH.

2.4.3. *Random access*

The procedure of random access is used for the following purposes:

– initial access by the mobile, to connect to the eNB;

– re-establishment of the connection after cutoff of the radioelectric link;

– handover during a session;

– re-establishment of the synchronization for uplink, if data need to be transmitted by the mobile and the timing advance value has expired;

– transmission of a scheduling request if no resources are attributed to the mobile.

Upon initial access, the mobile transmits the preamble over the physical channel PRACH. On receiving this message, the eNB is able to compute the time advance TA for the mobile (Figure 2.22).

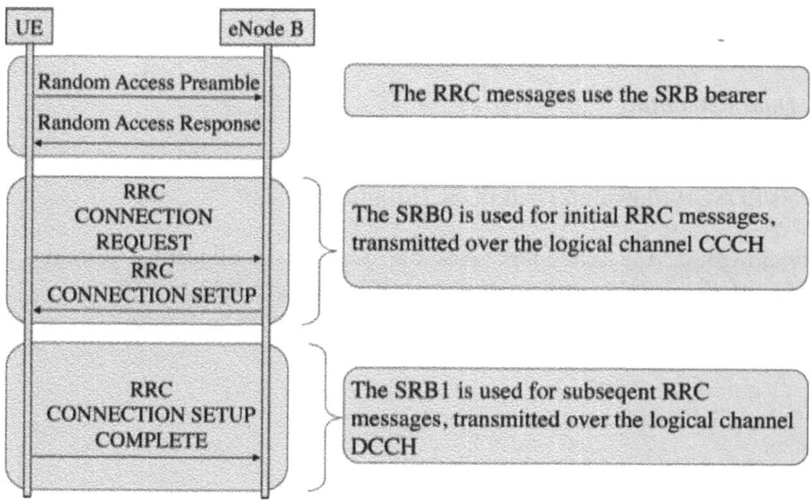

Figure 2.22. *Random access procedure*

The eNB responds to the mobile with a message MAC RAR RAPID containing the index of the preamble, the timing advance, the C-RNTI code temporarily attributed to the mobile and a grant to uplink transmission. The mobile has to process the physical channel PDCCH in order to find the RA-RNTI identifier, enabling it to retrieve the message MAC RAR RAPID multiplexed in the physical channel PDSCH (Figure 2.22).

If the mobile receives no response, it re-sends the preamble at an increased transmission power. It repeats the procedure in the absence of a response.

Multiple mobiles, using the same preamble at the same time on the same physical channel PRACH, may respond to the message MAC RAR RAPID, thereby giving rise to a contention. The mobiles send the message RRCConnectionRequest, indicating their S-TMSI code if the mobile is already attached or a random value otherwise (Figure 2.22).

The eNB responds with the message *RRCConnectionSetup* and the control element MAC containing the mobile ID (S-TMSI or random value), which resolves the contention. Mobiles which are not recognized have to repeat the random access procedure.

A mobile which is recognized by the message *RRCConnectionSetup* confirms receipt of the C-RNTI and responds with the message *RRCConnectionSetupComplete* (Figure 2.22).

2.4.4. *Data scheduling*

The eNB carries out data scheduling for the transmissions on the physical channels PDSCH (downlink) and PUSCH (uplink).

For downlink, the scheduler takes the decision to allocate a resource to the mobile on the basis of the following information (Figure 2.23):

– the category of the mobile, which determines the maximum datarate that it can receive;

– the available radio resources;

– the QCI associated with the radio bearer;

– the status of the queues of RLC data at the level of the eNB;

– the possible need for re-transmission indicated in the PUCCH;

– the quality of the signal received by the mobile (parameters CQI, PMI and RI reported by the mobile) in the physical channel PUCCH or PUSCH.

For uplink, scheduling can be activated by a request on the part of the mobile in the PUCCH channel. The scheduler then takes the decision to allocate a resource on the basis of the following criteria (Figure 2.24):

– the category of the mobile, which determines the maximum datarate that it can transmit;

– the available radio resources;

– the QCI associated with the radio bearer;

– the status of the queues of RLC data at the level of the mobile. This information is sent back in the PUSCH channel in a control element in the MAC layer;

– the possible need for re-transmission;

– the quality of the signal received by the eNB.

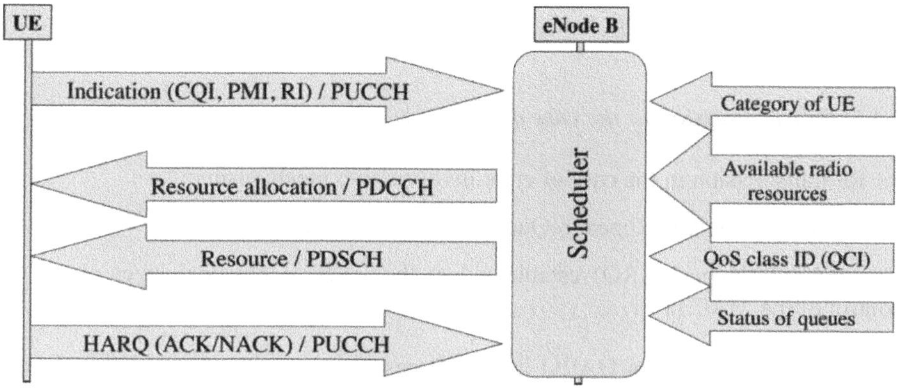

Figure 2.23. *Data scheduling for downlink*

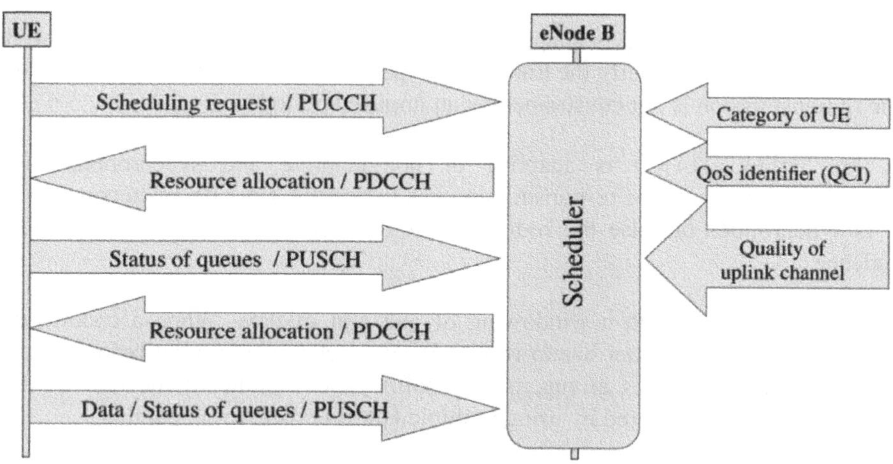

Figure 2.24. *Data scheduling for uplink*

Depending on the type of flow, scheduling will either be dynamic or semi-persistent. Dynamic scheduling consists of taking a decision every millisecond (the value of the TTI). This mode is apt for aperiodic or random flow such as telephone signaling. For this type of flow, the eNB assigns the mobile a C-RNTI code.

Semi-persistent scheduling reserves resources for a definite duration. This mode is apt for continuous flow such as voice or videophone. In this case, the resource requirements are predictable, which helps prevent an overload at the level of the PDCCH for resource allocation. For this type of flow, the eNB assigns the mobile an SPS-RNTI (Semi-Persistent Scheduling RNTI).

2.4.5. *Re-transmission in the case of error*

Re-transmission in the case of error involves two mechanisms:

– ARQ (Automatic Repeat reQuest) elaborated by the RLC layer;

– HARQ (Hybrid ARQ) established at the level of physical layer, under the control of the MAC layer.

ARQ takes over from HARQ if the re-transmissions at the level of the physical layer have failed. ARQ is applied only to the flows of traffic and signaling using AM in the RLC protocol. On the other hand, in view of its rapidity, HARQ is applied to flows using AM and UM in RLC.

For downlink, HARQ is adaptive and asynchronous. It is adaptive because the re-transmission may modify the transmission parameters. It is asynchronous because the re-transmission is not constrained by an imposed period.

For uplink, HARQ is adaptive or non-adaptive and synchronous. It is non-adaptive because the re-transmission has to use the same parameters as before. It is synchronous because the re-transmission has to be carried out at a definite cadence.

HARQ operates with a windowing of one unit of data. When a codeword is transmitted, the transmitter has to wait to receive acknowledgement before sending the next codeword. This arrangement penalizes the datarate of the mobile. This restriction can be combated by using multiple HARQ processes in parallel.

HARQ uses different redundancy versions during re-transmission. When a codeword is incorrectly received, it is kept by the receiver and then combined with the retransmitted codewords:

– in the case of Chase combining, the same codeword is re-transmitted. Linear combination with the previous codewords helps to improve the signal-to-noise ratio;

– in the case of incremental redundancy, re-transmission consists of sending additional redundancy bits produced by the turbo-code.

2.4.5.1. *Re-transmission for downlink*

Figure 2.25 illustrates the re-transmission for downlink of a codeword when the mobile has an allocation for voice data and for telephone signaling.

At time T0, the mobile receives, over the PDSCH, a transport block containing the voice data. It should be noted that this block is not notified over the PDCCH channel, because this is a case of semi-persistent scheduling. The allocation is associated with the following information:

– NDI (New Data Indicator), which differentiates the transmission of a new data block from a re-transmission;

– the number of the process (HARQ1);

– the redundancy version used for the transport block (RV0).

The mobile has detected residual errors in the received block. It sends a NACK response to the eNB over the PUCCH channel at T0+4 ms later.

The eNB re-sends the block over the physical channel PDSCH. The re-transmission takes place asynchronously. The transport block is notified over the PDCCH by the code SPS-RNTI, the number of the process HARQ1, the NDI with the value of ONE because it is a re-transmission, and the redundancy version RV1.

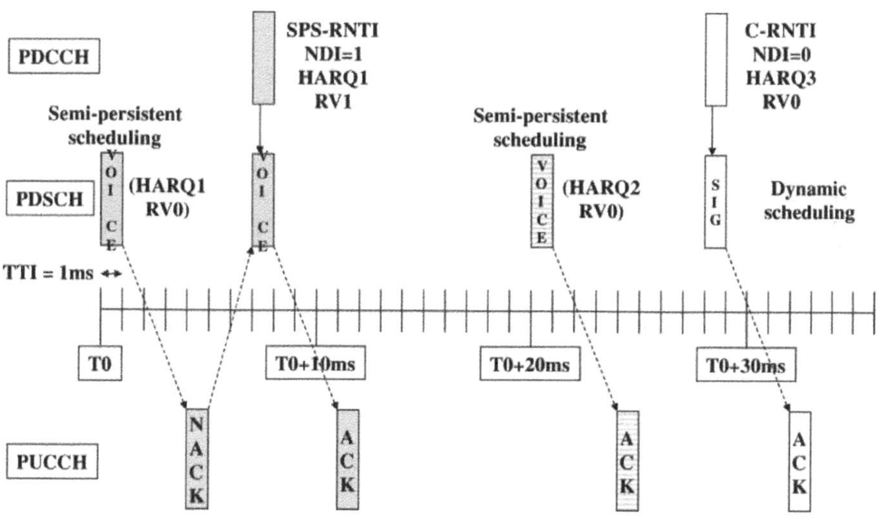

Figure 2.25. *Re-transmission for downlink*

If the mobile is successful in decoding the codeword, it responds with an ACK to the eNB over the PDCCH, 4 ms later.

At T0+20 ms, the mobile receives a new transport block containing voice data, associated with the process HARQ2. If the block is correctly decoded, the mobile responds with an ACK to the eNB over the PUCCH at T0+24 ms.

If the eNB has to transmit a transport block containing telephone signaling, it notifies the mobile of this over the PDCCH with the code C-RNTI, the process number HARQ3, the NDI with the value ZERO and the redundancy version (RV0). If the mobile is unable to decode the received transport block, the re-transmission mechanism is applied.

2.4.5.2. *Re-transmission for uplink*

Figure 2.26 illustrates the uplink re-transmission of a transport block when the mobile has an allocation for voice data and for telephone signaling. In the example given, the HARQ mechanism is non-adaptive for voice data and adaptive for telephone signaling.

At T0, the mobile transmits a transport block containing voice data over the physical channel PUSCH. The transport block is handled by the process HARQ1. This allocation is not indicated by the PDCCH channel because this is a case of semi-persistent scheduling.

If the eNB is unsuccessful in decoding the transport block, it sends a NACK message over the PHICH, at T0+4 ms.

The mobile has to re-send the transport block, having received the NACK. The re-transmission is non-adaptive because the mobile has not received a dynamic allocation on the TTI associated with the HARQ process. The mobile chooses the redundancy version independently, according to an established order.

Upon receipt of an allocation associated with the C-RNTI code over the PDCCH channel, the mobile transmits the telephone signaling over the PUSCH, 4 ms later.

If, 4 ms later, the mobile receives a NACK over the PHICH, and receives over the PDCCH a dynamic allocation on the TTI associated with the HARQ process, it re-transmits the transport block 4 ms later with the transmission characteristics indicated in the PDCCH channel.

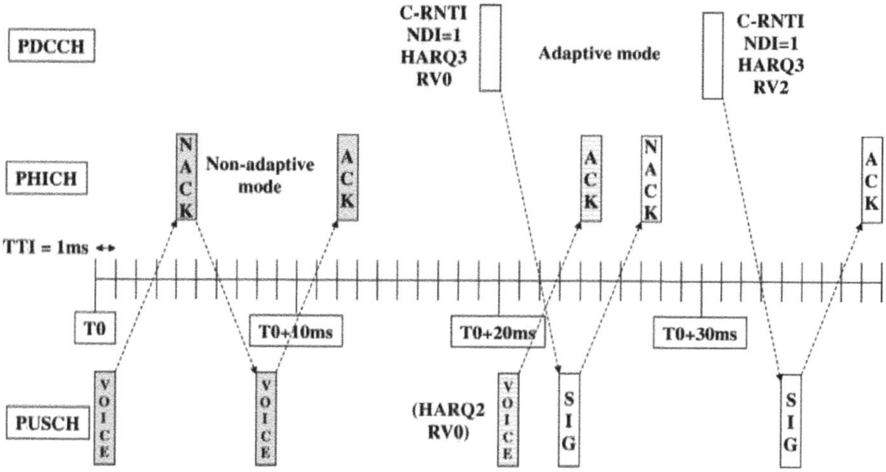

Figure 2.26. *Re-transmission for uplink*

Chapter 3

The CSFB Function

3.1. Reminder about NGN

3.1.1. *Architecture of NGN*

The NGN (Next Generation Network) is the core of the GSM (Global System for Mobile) network and UMTS (Universal Mobile Telecommunications System). It operates in CS (Circuit Service) mode.

The NGN is made up of two entities (Figure 3.1):

– firstly, the MSC (Mobile-services Switching Centre) Server or GMSC (Gateway MSC) Server;

– secondly, the MGW (Multimedia Gateway) and SGW (Signaling Gateway).

Interconnection between the entities forming the NGN takes place over an IP (Internet Protocol) network.

The MSC Server performs the functions of call management and mobility management. It completes the signaling exchanged with the following entities (Figure 3.1):

– the mobile UE (User Equipment) for CM (Call Management) or MM (Mobility Management) signaling;

– the access networks BSS (Base Station Sub-system), for BSSMAP (BSS Management Application Part) signaling and UTRAN (Universal Terrestrial Radio Access Network) for RANAP (Radio Access Network Application Part) signaling;

– the third-party networks PSTN (Public Switched Telephone Network) or PLMN (Public Land Mobile Network) for ISUP (ISDN User Part) signaling.

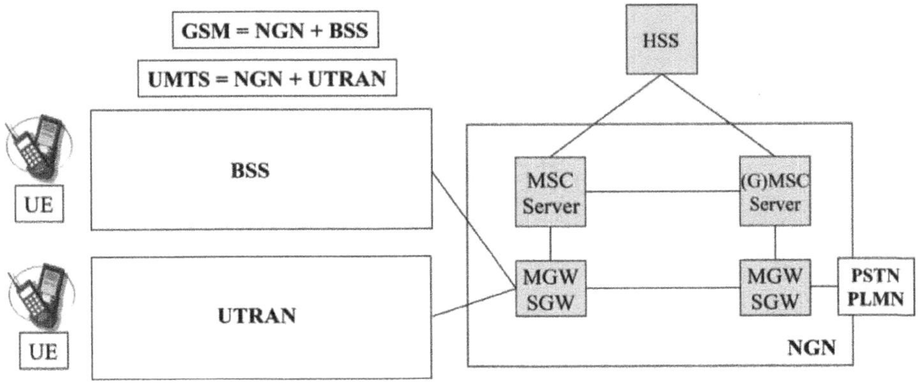

Figure 3.1. *Architecture of the NGN*

BSSAP signaling transports, on the one hand, the CM and MM signaling exchanged between the MSC Server and the UE, and, on the other hand, the BSSMAP signaling exchanged between the access network BSS and the MSC Server.

RANAP signaling carries the CM and MM signaling exchanged between the MSC Server and the UE. It also contains the messages exchanged between the access network UTRAN and the MSC Server.

The MSC Server can control many different MGWs/SGWs, but under no circumstances is it situated on the path of the media. It merely processes the signaling.

The GMSC Server provides the specific function corresponding to the handling of an incoming telephone call from the third-party network PSTN or PLMN.

The MSC Server takes care of the establishment, maintenance and release of connections in the MGW. A connection represents an association between an input end-point (e.g. the interface with the access networks or third-party networks) and an output end-point (e.g. the interface with the IP network) and vice versa (Figure 3.2).

The MGW handles the conversion of protocols relating to the flows of voice data between the two end-points. It also applies treatment, such as transcoding (modification of the type of codec between the two end-points), echo cancelling and generation of tones and announcements.

The SGW performs the conversion of transport protocols relating to the signaling exchanged between the MSC Server on the one hand and the access network and third-party networks on the other.

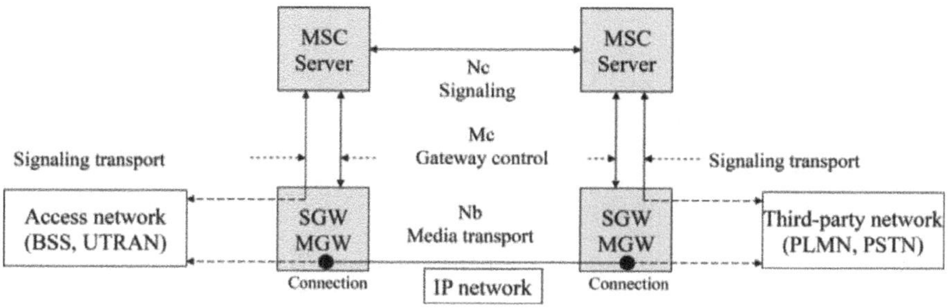

Figure 3.2. *The interfaces in the NGN*

Different interfaces are defined between the entities making up the NGN (Figure 3.2):

– the Mc interface between the MGW and MSC Server. This interface carries H.248 or MeGaCo (Media Gateway Controller) signaling for the control of the MGWs by the MSC Server;

– the Nc interface between the MSC Servers. This interface supports the BICC (Bearer Independent Call Control) signaling exchanged between the MSC Server entities. This signaling developed from the protocol ISUP, to which it is fundamentally similar with regard to the basic call procedures and the functions for supplementary telephone services;

– the Nb interface between the MGWs. This interface supports the transport of the voice data over the IP network.

The interface between the SGW and MSC Server, which forwards the signaling exchanged between the MSC Server on the one hand, and the access networks or third-party networks on the other, is unnamed.

3.1.2. *Signaling transport*

The model SS7 (Signaling System) is applied at the interfaces with the access network BSS, and the interfaces with the third-party networks. The model SIGTRAN (Signaling Transport over IP) is used on the internal interfaces of the

NGN, between the SGW and MSC Server entities. Conversion between the SS7 and SIGTRAN model is performed by the SGW (Figure 3.3).

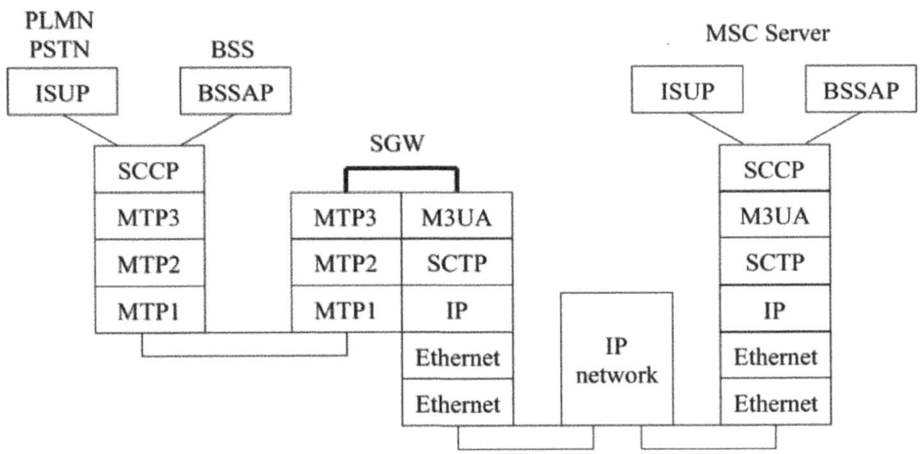

Figure 3.3. *Transport of signaling*

The SS7 model defines a data structure in layers, facilitating signaling transport in a time-switched digital network.

The protocol ISUP, corresponding to the application layer, defines the procedures used to establish, manage and release the circuits which transport the voice signals:

– IAM (Initial Address Message). This message is transmitted between digital switches to notify the request for establishment of a communication. It contains the telephone numbers of the caller and the callee;

– ACM (Address Complete Message). This message is returned by the destination switch to indicate that the ringing has been activated;

– ANM (Answer Message). This message is transmitted by the destination switch to indicate that the callee has picked up the telephone;

– REL (Release) message. This message is sent to release the resources when a user hangs up the telephone;

– RLC (Release Complete) message. This message is transmitted to acknowledge the REL message.

The SCCP (Signaling Connection Control Part) protocol, corresponding to the transport layer, provides the functions of end-to-end control between two end-points of SS7 signaling.

The MTP (Message Transfer Part) protocol transfers the messages. It is made up of three layers – MTP1, MTP2 and MTP3 – respectively corresponding to the physical layer, the data link layer and the network layer.

SIGTRAN defines two protocol layers adapting the transport of SS7 signaling over an IP network (Figure 3.3).

The SCTP (Stream Control Transmission Protocol) maintains the same characteristics of TCP (Transmission Control Protocol) and adds functions such as the multiplexing of several applications during the same session.

The protocol M3UA (MTP 3 User Adaptation) adapts the signaling protocols exchanged with the access networks or the third-party networks. M3UA supports the primitives between the MTP3 and the upper layer SCCP, in order to conceal from the MSC Server the fact that the MTP3 protocol terminates at the level of the SGW.

At the interface with the UTRAN, the structure of the SS7 data signaling is adapted to the use of ATM (Asynchronous Transfer Mode) as a transport protocol.

3.1.3. *Transport of voice data*

The MGW performs a conversion of the structure of the voice data between the IP network and the access networks or third-party networks (Figure 3.4).

At the interface with the access network BSS or with the third-party networks, the structure TDM (Time Division Multiplexing) directly multiplexes the voice data issuing from the G.711 codec in a timeframe G.704. The layer G.703 corresponds to the code HDB3 (High Density Binary) in the physical layer (Figure 3.4).

At the interface with the access network UTRAN, the voice data from the AMR (Adaptive Multi-Rate) codec are encapsulated by the protocols AAL2 (ATM Adaptation Layer), ATM and SDH (Synchronous Digital Hierarchy) (Figure 3.4).

At the interface of the third-party networks, the MGW also converts the format of the AMR codec used in the access network UTRAN into a G.711 codec, normalized for interconnection with the PLMN and PSTN (Figure 3.4).

At the interface of the IP network, the voice data are encapsulated by RTP (Real-Time Transport Protocol), UDP (User Datagram Protocol), IP (Internet Protocol) and Ethernet (Figure 3.4).

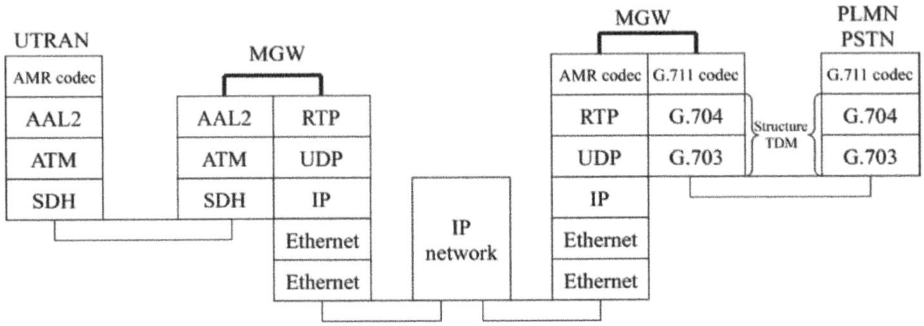

Figure 3.4. *Transport of voice data*

3.2. The CSFB function

The function CSFB (CS FallBack) enables a UE attached to the EPS (Evolved Packet System) network to use the GSM and UMTS networks to establish a telephone call. This function must be provided when mobiles can access a telephone service and the architecture IMS (IP Multimedia Subsystem) is not available.

CSFB is implemented by the creation of the SGs interface between, on the one hand, the MSC Server of the GSM and UMTS networks and, on the other, the MME (Mobility Management Entity) of the EPS network (Figure 3.5).

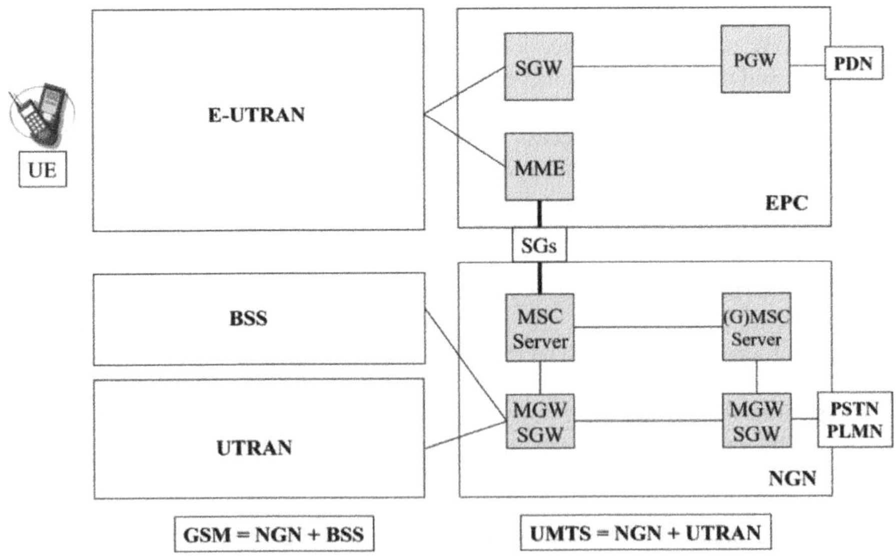

Figure 3.5. *SGs interface – connection between the networks*

The functions relating to the SGs interface are derived from the mechanisms of the Gs interface available between the MSC Server and the SGSN (Service GPRS Support Node) of the GPRS (General Packet Radio Service) mobile network. The MME appears, from the point of view of the MSC Server, as a BSS access network.

The protocol structure of the data exchanged over the SGs interface is shown in Figure 3.6.

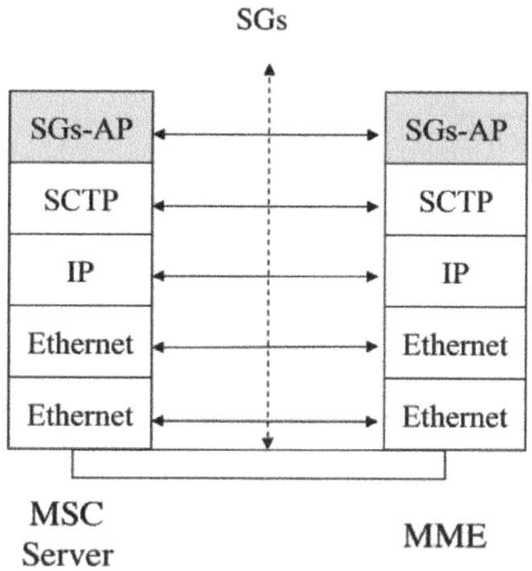

Figure 3.6. *SGs interface – protocol architecture*

The SGs-AP messages are used to fulfill the following functions:

– attachment of the mobile to the MSC Server;

– update of the mobile's Location Area Identifier (LAI) with the MSC Server;

– transmission of paging by the MSC Server to alert the mobile of an incoming telephone call.

3.3. Procedures

3.3.1. *Attachment*

When the UE initiates the procedure of attachment to the EPS network, it indicates in the message EMM ATTACH REQUEST that it has the capacity to handle the CSFB function (Figure 3.7). This message also contains the value of its TAI (Tracking Area Identifier) and IMSI (International Mobile Subscriber Identity).

The MME derives the value of the LAI on the basis of the TAI, and then derives the identity of the MSC Server from the LAI.

The MME sends the MSC Server the message SGs-AP IMSI ATTACH REQUEST, containing the value of the LAI and the IMSI (Figure 3.7).

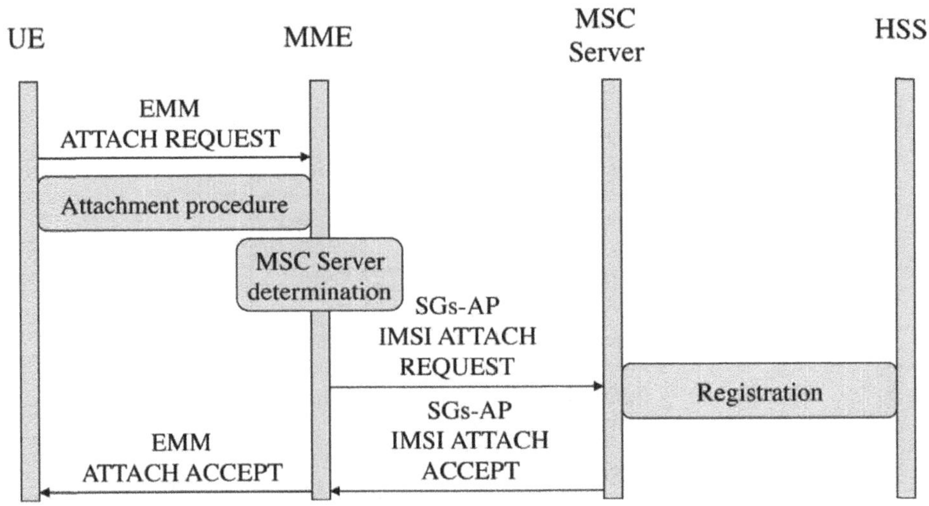

Figure 3.7. *Attachment*

The MSC Server registers the mobile with the HSS (Home Subscriber Server) and responds to the MME with the message SGs-AP IMSI ATTACH ACCEPT (Figure 3.7). This message contains the TMSI (Temporary Mobile Subscriber Identity).

This temporary identity is transferred to the mobile in the message EMM ATTACH ACCEPT (Figure 3.7). This message also contains the GUTI (Globally Unique Temporary Identity) assigned by the MME.

3.3.2. *Tracking area update*

When the mobile detects a change in its TAI, it sends the message EMM TRACKING AREA UPDATE REQUEST, whose destination might be a new MME. This message contains the temporary identities TMSI and GUTI (Figure 3.8).

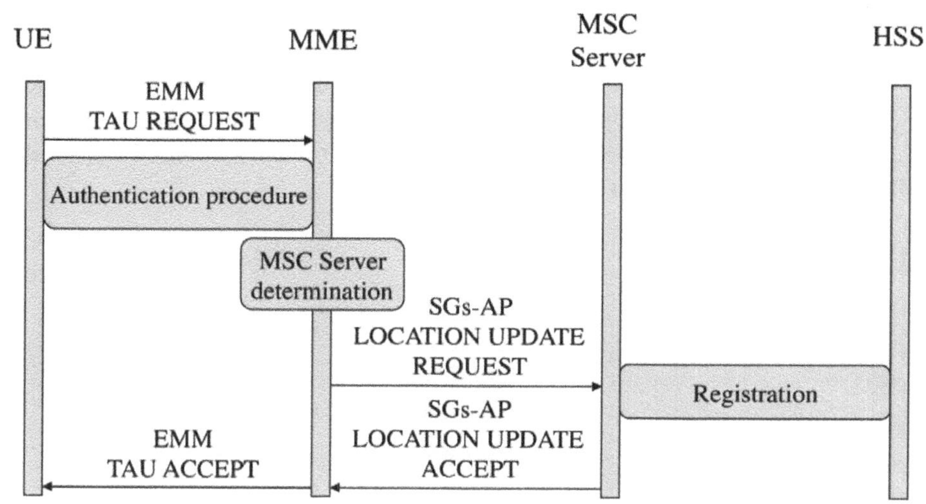

Figure 3.8. *Location update*

The MME will possibly derive the new value of the LAI from the new TAI, and then derives the identity of the MSC Server from the new LAI.

An LAI zone can encapsulate several TAI zones. A change of TAI does not necessarily imply a change of LAI. In addition, a TAI cannot and must not overlap between more than one LAI – it must be a sub-set of a lone LAI.

The MME sends the MSC Server the message SGs-AP LOCATION UPDATE REQUEST containing the new value of the LAI and the TMSI of the mobile (Figure 3.8).

The MSC Server updates the location of the mobile with the HSS if the new LAI corresponds to a new MSC Server.

The MSC Server responds to the MME with the message SGs-AP LOCATION UPDATE ACCEPT (Figure 3.8). This message contains the new TMSI of the mobile.

The new TMSI and GUTI are transferred to the mobile in the message EMM TRACKING AREA UPDATE ACCEPT (Figure 3.8).

3.3.3. *Outgoing call*

When the mobile is active on the EPS network, it is connected to the eNB (evolved Node B) and attached to the MME. If it wishes to establish a telephone call, it must then send the message EMM EXTENDED SERVICE REQUEST to the MME to initiate the function CSFB (Figure 3.9).

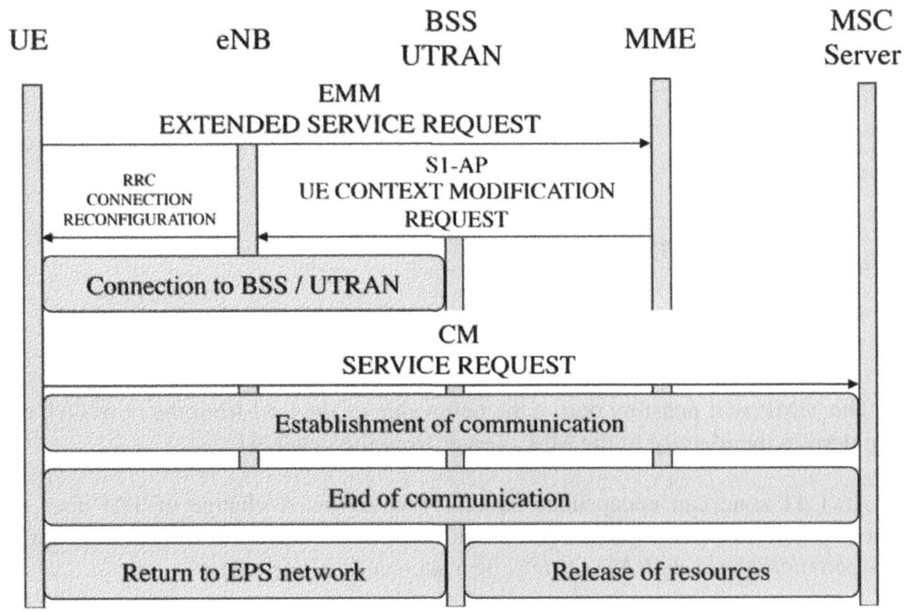

Figure 3.9. *Outgoing call*

This message is carried by the message RRC *ULInformationTransfer* over the radioelectric interface and the message S1-AP UPLINK NAS TRANSPORT over the S1-MME interface.

The MME sends the message S1-AP UE CONTEXT MODIFICATION REQUEST to the eNB to alert it to the fact that the mobile has initiated the CSFB procedure. The eNB triggers the handover by sending the message *RRCConnection Reconfiguration* so that the mobile is transferred to the GSM or UMTS network (Figure 3.9).

The eNB can, optionally, request a measurement report from the mobile in order to determine the target cell of the access network BSS or UTRAN.

As the mobile is active, the eNB initiates the handover for the current operation in PS (Packet Service) mode on the EPS network. If the target access network is a BSS, the mobile suspends service with the SGSN, unless the function DTM (Dual Transfer Mode) is supported.

The stopping of service in PS mode causes the deactivation of those bearers whose datarate was guaranteed, and the suspension of those bearers whose datarate was not guaranteed.

When the mobile is connected to the access network BSS or UTRAN, it begins the procedure of initialization of the telephone call by sending the message CM SERVICE REQUEST to the MSC Server (Figure 3.9). This message indicates that the call is being made following the initiation of CSFB.

If the access network BSS or UTRAN has been notified by the MSC Server that the telephone call has been initiated by the function CSFB, it has to transfer the mobile back to the EPS network when the call ends.

When the mobile returns to the EPS network, and if the service in PS mode has been suspended, the mobile sends the message EMM EXTENDED SERVICE REQUEST to the MME so it will reactivate the bearers.

If the mobile was in idle mode when the CSFB function was initiated, it must first establish the connection with the eNB before sending the message EMM TRACKING AREA UPDATE REQUEST. Upon receiving this message, the MME begins the procedure of mutual authentication.

3.3.4. *Incoming call*

An incoming call corresponds to the receipt of the message BICC IAM by the MSC Server, this message is sent from the GMSC Server once it receives the call from the PSTN or PLMN (Figure 3.10).

The MSC Server sends the MME the message SGs-AP PAGING REQUEST, indicating the mobile's IMSI or temporary TMSI (Figure 3.10).

If the mobile is active on the EPS network, the MME sends it the message EMM CS PAGING NOTIFICATION (Figure 3.10). The next step begins with the mobile sending the message EMM EXTENDED SERVICE REQUEST to the MME. The rest of the procedure corresponds to that described for an outgoing call.

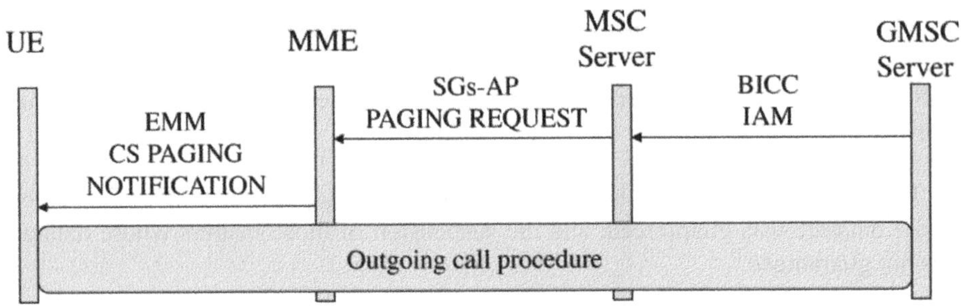

Figure 3.10. *Incoming call – the mobile is connected*

If the mobile is in idle mode, and if the MME receives the TMSI in the message SGs-AP PAGING REQUEST, the MSC Server uses the S-TMSI (Shortened-TMSI) to page the mobile. The S-TMSI is a subset of the GUTI.

If the IMSI is used in the message SGs-AP PAGING REQUEST, the MME uses the IMSI to page the mobile.

The MME sends all the eNBs in the TAI the message S1-AP PAGING. Every eNB in turn broadcasts the message RRC *Paging* to bring the mobile out of idle mode (Figure 3.11).

The mobile begins the connection procedure for establishment of the SRB (Signaling Radio Bearer), and then sends the message EMM EXTENDED SERVICE REQUEST to the MME.

The MME then sends the message SGs-AP SERVICE REQUEST to the MSC Server, which stops the re-transmission of the message SGs-AP PAGING REQUEST. The rest of the procedure involves the same exchanges as described above for the outgoing call (Figure 3.11).

As with outgoing calls, the BSS or UTRAN access network can transfer the mobile back to the EPS network when the telephone call is ended.

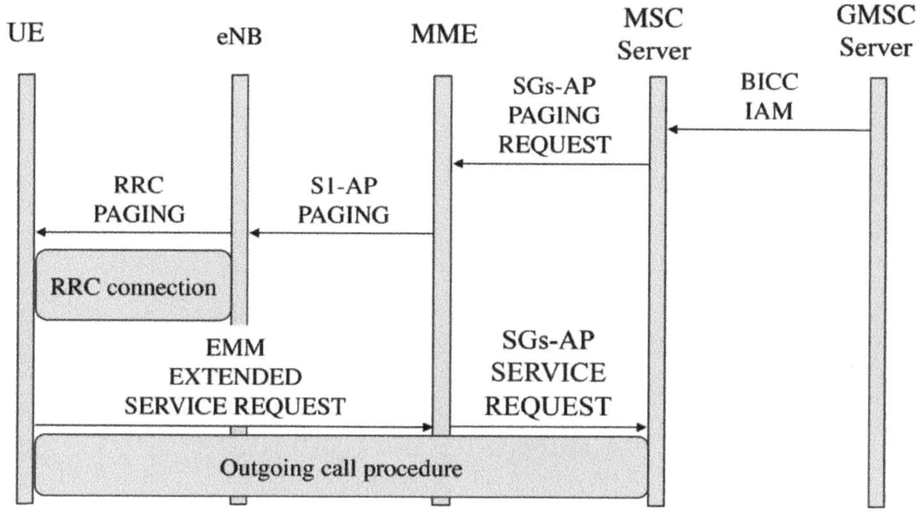

Figure 3.11. *Incoming call – the mobile is in idle mode*

Chapter 4

SIP and SDP Protocols

4.1. Entities

The UA (User Agent) is a logical entity located in a terminal (hardphone or softphone), in a gateway or in an application (e.g. voicemail).

The UAC (User Agent Client) is the logical entity which creates a request and establishes a transaction. The role of UAC only lasts for the duration of that transaction. If, subsequently, the UA receives a request, it assumes the role of UAS (User Agent Server) to deal with this new transaction.

A UAS is the logical entity which generates a response to the received request. The response accepts, rejects or redirects the request. This role only lasts for the duration of that transaction. If, subsequently, the UA generates a request, it assumes the role of a UAC to deal with this new transaction.

The two basic requests transmitted by the UA are REGISTER for the registration of the UA, and INVITE for the establishment of a session.

The REGISTRAR is the entity which accepts the REGISTER requests. It places the IP address and the URI (Uniform Resource Identifier) that it receives in the REGISTER request from the UA in a database – the LOCATION SERVER (Figure 4.1).

The PROXY SERVER is an intermediary entity which takes care of the routing of the INVITE request, of subsequent requests and of the responses. It ensures that the request is sent either to another PROXY SERVER or to the target user.

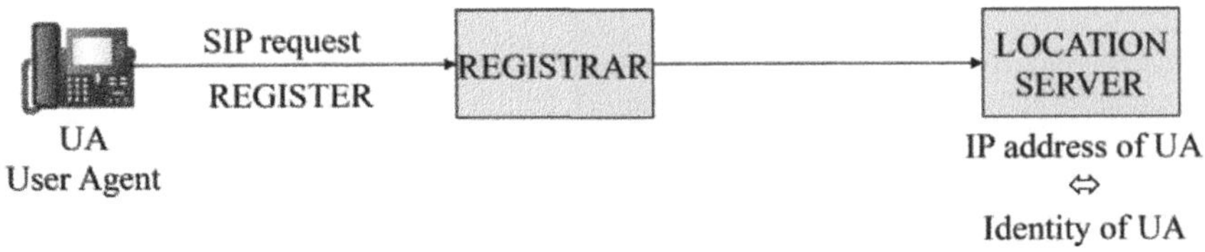

Figure 4.1. *Registration*

The PROXY SERVER interprets and, if necessary, re-writes specific parts of an SIP (Session Information Protocol) message before re-transmitting it. It can also carry out an authentication of the UA that generated the request (Figure 4.2).

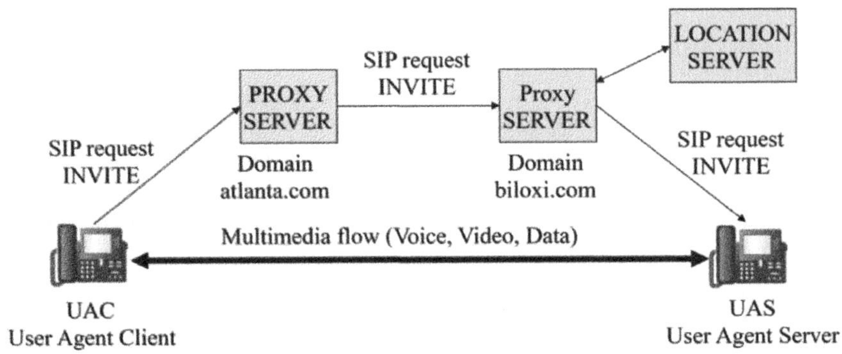

Figure 4.2. *Establishment of session*

The LOCATION SERVER is used by a PROXY SERVER to obtain information about the location (IP address) of the UAS being called (Figure 4.2).

The REDIRECT SERVER is a redirection entity which generates a response to the requests received in order to direct the call from the UAC to a different target.

The B2BUA (Back-to-Back User Agent) is a logical entity which receives a request and treats it as a UAS. In order to determine the appropriate response, it acts as a UAC and generates requests directed at the target. The B2BUA can translate from IPv4 to IPv6 protocols, or convert IPv4 private addresses into public addresses.

4.2. Identities

A UA has an identity called a URI. This identity may take one of three different forms: SIP URI, SIPS URI or TEL URI.

The SIP URI is similar in form to an e-mail address, typically containing a username (e.g. Alice or Bob) and a domain name (e.g. atlanta.com or biloxi.com). Thus, we obtain the following two URIs:

`sip:alice@atlanta.com`

`sip:bob@biloxi.com`

The SIPS URI `sips:alice@atlanta.com` is employed when the signaling is secured by the TLS (Transport Layer Security) protocol it is used to transport all of the SIP messages from the UAC to its PROXY SERVER, between PROXY SERVERS or between the PROXY SERVER and the UAS.

The TEL URI is represented by a telephone number. It has the form `tel:+33123456789`. The character + indicates that this is a worldwide number, and is followed by the country code. The caller's PROXY SERVER has to translate the TEL URI into a SIP URI (this is ENUM resolution) in order to find the next PROXY SERVER, to which it transfers the SIP message.

4.3. Structure of SIP

SIP is a control protocol which can establish, modify and terminate multimedia sessions. Media can be added to or removed from an existing session.

SIP is based on a request/response pair such as HTTP (Hypertext Transfer Protocol). Each transaction consists of a request, which uses a particular method, and one or more responses.

4.3.1. *Requests*

The request begins with a line containing the method, a URI and the version of the protocol. It then contains headers which are discussed in section 4.2.3.

`INVITE sip:bob@biloxi.com SIP/2.0`

4.3.1.1. *REGISTER method*

The REGISTER method is used by a UA to notify the REGISTRAR of the correspondence between the IP address of the UA and its URI. This correspondence is necessary for incoming calls.

The use of the URI of the REGISTER request, and of the headers `To` and `From`, is slightly different to that of other requests. The URI contains only the domain of the REGISTRAR, with no part relating to the user. The header `To` contains the URI of the UA which needs to be registered. The header `From` contains the URI of the entity performing the registration. This entity is generally identical to that of the header `To`.

A UA may receive a redirect (`3xx`) or failure (`4xx`) response, whose header `Contact` contains the place where the registrations need to be sent.

4.3.1.2. *INVITE method*

The INVITE method is used by a UAS to establish a dialog or a session. The definitive (positive or negative) responses need to be acknowledged by the ACK request.

The INVITE request may contain a message body describing the media that the UAC wants to establish. If this description is absent, it needs to be added to the ACK request.

The response `200 OK` contains the description of the media that the UAS wants to establish. The media are established with the response `200 OK` (if the INVITE request contains the description of the media) or with the ACK request (otherwise).

A successful INVITE request establishes a dialog between two UAs, which continues until a BYE request is sent by one of the parties to terminate the session. The dialog is identified by the header `Call-ID` and the parameter `tag` of the headers `To` (parameter specified by the caller) and `From` (parameter specified by the callee). The headers `To` and `From` are respectively specified with the identities of the callee and the caller.

Within the dialog, it is possible for each UA to transmit one or more requests, with each request initializing a transaction. It is possible to transmit a new INVITE request (reINVITE) within a dialog, for an established session, in order to update the characteristics of the media. If the reINVITE request is declined, the session continues with the previous characteristics.

4.3.1.3. *ACK method*

The ACK method is used to acknowledge a definitive response (`2xx`, `3xx`, `4xx`, `5xx`, `6xx`) to the INVITE request. The ACK request may contain a message body describing the media if this description is not given in the INVITE request.

4.3.1.4. *BYE method*

The BYE method is used to terminate an established session. A session is considered to be established when the response `2xx` is received following the INVITE request.

The BYE request is transmitted by either of the UAs participating in the session. The UAS sends the response back `2xx` if the dialog is known, or else sends the response `481 Dialog/Transaction Does Not Exist`.

4.3.1.5. *CANCEL method*

The CANCEL method is used to terminate a session which has not been successful. It is generated when a provisional response `1xx` has been received, but not a definitive response.

Upon receiving a CANCEL request, the PROXY SERVER transmits the request to the next hop (a PROXY SERVER or a UA) and confirms the cancellation directly to the source with the response `200 OK`. A UAS receiving the CANCEL request confirms cancellation with the response `200 OK` and terminates the dialog initiated by the INVITE request with a definitive negative response `487 Request Terminated`.

4.3.1.6. *PRACK method*

The PRACK method is used to acknowledge receipt of a provisional response (`1xx`), with the exception of the response `100 Trying`.

The PRACK request is transmitted when the provisional response received contains the headers `CSeq` and `Rseq`. The PRACK request must include the header `RAck` containing the values of the headers `CSeq` and `Rseq` of the provisional response received.

The provisional response is transmitted at the expiration of a timer until the reception of the PRACK request.

The PRACK request is transmitted at the expiration of a timer until the reception of a `200 OK` response.

4.3.1.7. *UPDATE method*

The UPDATE method modifies the characteristics of the media in a session which has not yet been successfully established. If the session has successfully been established (the INVITE request has received a `2xx` response), the modification of the characteristics of the media is negotiated by the INVITE method (reINVITE).

4.3.1.8. *SUBSCRIBE method*

The SUBSCRIBE method is used when a UA wishes to subscribe to a service whereby he would receive event notifications. The type of event is described in the header `Event`.

The entity which accepts the subscription returns the response `200 OK` containing the duration of the subscription in the header `Expires`. The subscription has to be renewed by transmitting a new SUBSCRIBE request. In the absence of renewal, the subscription is automatically terminated.

4.3.1.9. *NOTIFY method*

The NOTIFY method enables an entity to notify the occurrence of an event.

This entity needs to receive a response `200 OK` which gives the assurance that the request has been properly received. Receipt of the response `481 Dialog/Transaction Does Not Exist` automatically terminates the subscription.

4.3.1.10. *REFER method*

The REFER method can be used to transfer media established between two UAs (e.g. Alice and Bob) to someone else. The REFER request is sent by Alice (transferor) to Carol to resume communication. The header `Refer-to` of the request contains the SIP URI of Bob (the transferee). It should be noted that in this scenario, communication is not established between Alice and Carol.

In a second scenario, Alice establishes an additional communication with Carol. Alice then sends the REFER request to Bob to transfer the communication that she has established with Carol. When Bob notifies her that the transfer has been successful, Alice releases the communication established with Bob.

4.3.1.11. *MESSAGE method*

The MESSAGE method is used to transmit small messages, contained in a message body, between two UAs. This message is acknowledged by a response `200 OK`. The response must not contain the message body. If the recipient wishes to respond, he must in turn generate a new MESSAGE request.

4.3.2. *Responses*

The response begins with a line containing the version of the protocol, followed by a code for the type of response and a textual description of the code. It contains the headers described in section 4.2.3.

```
SIP/2.0 200 OK
```

The different types of responses are made explicit in Table 4.1.

Type of response	Designation
`1xx`	Provisional response.
`2xx`	Definitive positive response.
`3xx`	Definitive redirect response.
`4xx`	Definitive negative response, error due to client.
`5xx`	Definitive negative response, error due to network.
`6xx`	Definitive negative response, global error.

Table 4.1. *Types of responses*

4.3.2.1. *1xx responses*

4.3.2.1.1. 100 Trying

The response `100 Trying` is generated by the PROXY SERVER to alert the sender of the request (the UAC) that the SIP message has been received. It does not contain the header `To`.

4.3.2.1.2. 180 Ringing

The response `180 Ringing` is used by the callee (the UAS) to tell the caller (the UAC) that the INVITE request has been received and that the callee has been alerted to an incoming call by a ringing tone.

4.3.2.1.3. 181 Call Is Being Forwarded

The response `181 Call Is Being Forwarded` is transmitted by the callee (the UAS) to tell the caller (the UAC) that its call has been transferred to a different recipient.

4.3.2.1.4. 183 Session Progress

The response `183 Session Progress` is transmitted by the callee (the UAS) to inform the caller (the UAC) that its call is being processed.

4.3.2.2. *2xx responses*

4.3.2.2.1. 200 OK

The response `200 OK` has two uses. When it is a response to an INVITE request, it contains the message body describing the media put in place by the UAS. For other requests, the response acknowledges receipt of the request.

4.3.2.2.2. 202 Accepted

The response `202 Accepted` indicates that the callee has indeed received the request, requiring a different treatment thereafter. This response is associated with the SUBSCRIBE and REFER requests. The completeness of the subscription request or communication transfer request is finalized by the NOTIFY request.

4.3.2.3. *3xx responses*

Table 4.2 outlines the `3xx` responses.

Response	Description
`300 Multiple Choices`	The redirect indicates multiple contacts, the order of which is significant.
`301 Moved Permanently`	The redirect is permanent.
`302 Moved Temporarily`	The redirect is temporary.
`305 Use Proxy`	The redirect is performed to a PROXY SERVER.
`380 Alternative Service`	The redirect points to a different entity (e.g. voicemail service).

Table 4.2. *3xx responses*

4.3.2.4. *4xx responses*

4.3.2.4.1. 401 Unauthorized

The response `401 Unauthorized` indicates that the request requires authentication from the UA. It is used by the REGISTRAR on receiving a REGISTER request.

The response includes the header `WWW-Authenticate`, which contains a challenge that the UA uses to compute the authentication data.

4.3.2.4.2. 407 Proxy Authentication Required

The response `407 Proxy Authentication Required` also indicates that the request requires authentication from the UA. It is used by the PROXY SERVER on receiving an INVITE request.

The response includes the header `Proxy-Authenticate` which contains a challenge that the UA uses to compute the authentication data.

4.3.2.4.3. 486 Busy Here

The response `486 Busy Here` indicates that the callee is busy, so the communication cannot be established. The response terminates the dialog initialized by the INVITE request.

4.3.2.4.4. 487 Request Terminated

The response `487 Request Terminated` is transmitted by the UA when it receives a CANCEL request. It responds with a `200 OK` to the CANCEL request and with a `487 Request Terminated` to end the dialog initialized by the INVITE request.

4.3.2.5. *5xx responses*

Table 4.3 outlines the `5xx` responses.

Response	Description
`500 Server Internal Error`	An internal error has occurred on the PROXY SERVER or the REGISTRAR, so the request has to be repeated later.
`502 Bad Gateway`	The gateway to a different network has detected a fault.
`503 Service Unavailable`	The service is temporarily unavailable.
`504 Gateway Timeout`	The gateway to a different network cannot relay the request because a timer has run out.
`505 Version Not Supported`	The request is denied because of the version of the SIP.
`513 Message Too Large`	The request is denied because the size of the SIP message is too great.

Table 4.3. *5xx responses*

4.3.2.6. *6xx responses*

Table 4.4 outlines the `6xx` responses.

Response	Description
`600 Busy Everywhere`	The network is saturated.
`603 Decline`	The callee refuses the call.
`604 Does Not Exist Anywhere`	The SIP URI of the callee does not exist anywhere.
`606 Not Acceptable`	Some aspects of the session are not acceptable.

Table 4.4. *6xx-type responses*

4.3.3. *Headers*

4.3.3.1. *The header Via*

The header `Via` is used to record the itinerary of the request and thereby enable the response to be directed via a certain path. The UAC generating a request registers its own address in a `Via` header. The order of the `Via` headers is important, because the response has to follow the same path as the request, in the opposite direction.

The PROXY SERVERS relaying the request add a `Via` header containing their own address to the start of the list of `Via` headers. The UAS generating a response to the request copies all the `Via` headers and sends the response to the address indicated by the first header on the list.

Upon receiving a response, the PROXY SERVER verifies the first header on the list in order to ensure that it corresponds to its own address, then removes it and transfers the response to the address indicated in the next `Via` header.

The `Via` header contains the name of the protocol, its version, the type of transport protocol (SIP/2.0/UDP, SIP/2.0/TCP, etc.), and parameters such as `branch`.

The UAC and PROXY SERVER add a parameter `branch` which identifies the transaction in progress.

```
Via: SIP/2.0/UDP 192.0.2.201;branch=z9hG4bKnashds7
```

4.3.3.2. *The header From*

The header `From` indicates the source of the request. The UAC adds the parameter `tag` which takes part in the identification of the dialog.

```
From: Alice <sip:alice@atlanta.com>;tag=9fxced76sl
```

4.3.3.3. *The header To*

The header `To` indicates the destination of the request. The responses generated by the UAS contain this header with the addition of a parameter `tag`.

```
To: Bob <sip:bob@biloxi.com>;tag=3flal12sf
```

The value of the parameter `tag` in the header `To`, combined with that of the `tag` in the header `From` and that of the header `Call-ID`, identifies the dialog.

The header `To` is not used to route the SIP message. The URI of the request is used for this purpose.

4.3.3.4. *The header Call-ID*

The header `Call-ID` is part of the identification of the dialog between two UAs. `Call-ID` must be unique for a call.

The value of the header `Call-ID` is set by the UAC and is never modified by the PROXY SERVERS.

```
Call-ID: 3848276298220188511@atlanta.com
```

4.3.3.5. *The header CSeq*

The header `CSeq` contains a decimal number which increases with each request, with the exception of the requests CANCEL and ACK, which use the sequence number of the INVITE request to which they refer.

```
CSeq: 1 INVITE
```

4.3.3.6. *The header Contact*

The header `Contact` is used to communicate the IP address of the transmitter of the request or of the response. This address is cached for the routing of subsequent requests in a dialog.

The parameter `expires` indicates the validity time of the registration of the identity of the transmitter.

```
Contact: <sip:bob@192.0.2.201>;expires=3600
```

4.3.3.7. *The header Max-Forwards*

The header `Max-Forwards` indicates the maximum number of hops that a request can take. The value of this header is decreased by each PROXY SERVER that transfers the request.

```
Max-Forwards: 70
```

A PROXY SERVER which receives this header with a value of zero deletes the message and sends a response `483 Too Many Hops` to the source of the message.

4.3.3.8. *The header Content-Type*

The header `Content-Type` specifies the possible types of content for the message associated with the SIP message.

```
Content-Type: application/sdp
```

4.3.3.9. *The header Content-Length*

The header `Content-Length` indicates the number of bytes in the message body. A value of zero indicates that the message body is blank. The byte counter does not include the CRLF (Carriage Return Line Feed) character which separates the SIP message from the message body.

```
Content-Length: 151
```

4.3.3.10. *The header Route*

The header `Route` is used to convey information about the routing of the requests. Two modes of routing – strict routing and loose routing – are defined.

In strict routing mode, the request absolutely has to follow the itinerary set out in the list of `Route` headers. The PROXY SERVER has to use the address of the header `Route` to replace the URI of the request before transferring the request.

In loose routing mode, the request has to be forwarded via each PROXY SERVER identified in the list of headers `Route`, but can also be routed through other PROXY SERVERS. The PROXY SERVER must not modify the URI of the request.

```
Route: <sip: 192.0.2.111;lr>
```

4.3.3.11. *The header Record-Route*

The header `Record-Route` is used to force the routing through a PROXY SERVER for all subsequent requests between two UAs. In this header, the PROXY SERVER indicates its address during the initial request. The UAS copies the list of headers `Record-Route` in the response. The list is transmitted to the UAC, without being modified by the PROXY SERVERS.

The parameter `lr` indicates that the PROXY SERVERS support loose routing for the SIP messages.

```
Record-Route: <sip:192.0.2.111;lr>
```

4.3.3.12. *The header WWW-Authenticate*

The header `WWW-Authenticate` is used in the response `401 Unauthorized`. It contains a challenge of authentication (parameter `nonce`) of the UAC by the REGISTRAR.

```
WWW-Authenticate: Digest realm="atlanta.com",
qop="auth", nonce="ea9c8e88df84f1cec4341ae6cbe5a359",
opaque="", stale=FALSE, algorithm=MD5
```

4.3.3.13. *The header Authorization*

The header `Authorization` transports the authentication data (parameter `response`) of the UA to the REGISTRAR. It is sent after a response `401 Unauthorized` containing a challenge.

```
Authorization: Digest username="bob",
realm="atlanta.com"
nonce="ea9c8e88df84f1cec4341ae6cbe5a359", opaque="",
uri="sip:biloxi.com",
response="dfe56131d1958046689d83306477ecc"
```

4.3.3.14. *The header Proxy-Authenticate*

The header `Proxy-Authenticate` is used in the response `407 Proxy Authentication`. It contains the challenge for authentication of the UAC by the PROXY SERVER.

```
Proxy-Authenticate: Digest realm="atlanta.com",
qop="auth", nonce="f84f1cec41e6cbe5aea9c8e88d359",
opaque="", stale=FALSE, algorithm=MD5
```

4.3.3.15. *The header Proxy-Authorization*

The header `Proxy-Authorization` transports the authentication data of the UA to the PROXY SERVER. It is sent after a response `407 Proxy Authentication Required` containing a challenge.

```
Proxy-Authorization: Digest
username="alice",realm="atlanta.com",nonce="wf84f1ceczx4
1ae6cbe5aea9c8e88d359", opaque="",
uri="sip:bob@biloxi.com",
response="42ce3cef44b22f50c6a6071bc8"
```

4.3.3.16. *The header Refer-To*

The header `Refer-To` is used with the request REFER. It specifies the SIP URI to which the communication needs to be transferred.

```
Refer-To: Carol <sip:carol@chicago.com>
```

4.3.3.17. *The header Replaces*

When an entity has received the request REFER, it generates an INVITE request to effect the transfer of the communication. The header `Replaces` contains the identifiers of the dialog in question: header `Call-ID`, parameters `tag` of the headers `From` and `To`.

```
Replaces=1234540@192.168.118.3;to-tag=12345;from-
tag=3994>
```

4.3.3.18. *The header RSeq*

The header `RSeq` is used in `1xx`-type responses to an INVITE request. It contains a sequence number which identifies the response. It is introduced because the INVITE request contains the header `Supported`, which takes the value `rel100`.

```
RSeq: 12345
```

4.3.3.19. *The header RAck*

The header `RAck` is used in the PRACK request which validates the receipt of a `1xx`-type response. It contains the identifiers of the INVITE request (`CSeq`) and the response to be validated (`RSEq`).

```
RAck: 12345 1 INVITE
```

4.4. Description of the media

The SDP (Session Description Protocol) gives a description of the media flow for which the establishment of the session is implemented by the SIP. The SDP message constitutes the message body attached to the SIP message. It generally appears in the INVITE request and in the response `200 OK`.

The parameters which characterize the media flows are as follows:

– the type of media (audio, video, data);

– the transport protocol (e.g. RTP);

– the format of the media (e.g. the type of codec for voice and video data);

– the IP address to which the media need to be transmitted;

– the number of the destination port.

The SDP message is a set of lines of code in the format <type>=<value>. The field <type> contains a character (Table 4.5). The content of the field <value> depends on the type.

Field <type>	Description
v	Version of the protocol. The content of the field <value> is <0>
o	Identifier of the origin and of the session
s	Name of the session. The content of the field <value> is <->
c	Information about the connection
t	Activity time of the session. The content of the field <value> is <0 0>
m	Description of the media
a	Complementary attribute of the media

Table 4.5. *SDP message*

The field <value> corresponding to the field <o> contains the following parameters:

```
o=alice 2890844526 2890844526 IN IP4 192.0.2.101
```

– username (`alice`);

– session ID (`2890844526`). This value is fixed when the SDP message is associated with the SIP message;

– session version (`2890844526`). This value is fixed when the SDP message is associated with the SIP message;

– type of network (`IN`). The value of the field corresponds to the Internet Network;

– type of address (`IP4`);

– IP address of the machine that created the session (`192.0.2.101`).

The field <value> corresponding to the field <c> contains the following parameters:

```
c=IN IP4 192.0.2.101
```

– type of network (`IN`). The value of the field corresponds to the Internet Network;

– type of address (`IP4`);

– IP address to which the media have to be sent (`192.0.2.101`).

The field <value> corresponding to the field <m> contains the following parameters for the establishment of an audio flow:

```
m=audio 49172 RTP/AVP 0
```

– type of media (`audio`);

– the port number to which the audio flow has to be transmitted (`49172`). The value indicated is that of the RTP flow. The following value is used for RTCP messages;

– type of transport protocol (`RTP/AVP`);

– type of codec (`0`). This value is identical to that used in the RTP protocol to define the type of codec.

The field <value> corresponding to the field <a> contains the following parameters, which provide an additional description of the audio flow:

```
a=rtpmap:0 PCMU/8000
```

– type of codec (`0`). This value is identical to that used in the RTP protocol to define the type of codec;

– name of codec (`PCMU`);

– sampling frequency (`8000`).

4.5. Procedures

4.5.1. *Registration*

A UA registers with the REGISTRAR. It determines the destination IP address in the following way:

– the IP address is obtained by configuration, e.g. via a DHCP server;

– the IP address is the multicast address 224.0.1.75;

– the IP address is obtained by performing a DNS (Domain Name System) resolution on the basis of the domain name.

The URI of the REGISTER request corresponds to the name of the LOCATION SERVER with which Bob is registered.

The header `To` contains the URI to be registered. The header `From` contains the ID of the entity doing the registration. The header `Contact` contains the IP address which needs to be linked to the SIP URI of the header `To`.

The dialog is identified by the header `Call-ID` and the parameter `tag` of the header `From`.

```
REGISTER sip:biloxi.com SIP/2.0
Via: SIP/2.0/UDP 192.0.2.201;branch=z9hG4bKnashds7
Max-Forwards: 70
From: Bob <sip:bob@biloxi.com>;tag=a73kszlfl
To: Bob <sip:bob@biloxi.com>
Call-ID: 1j9FpLxk3uxtm8tn@biloxi.com
CSeq: 1 REGISTER
Contact: <sip:bob@ 192.0.2.201>
Content-Length: 0
```

The REGISTRAR has to authenticate Bob. Therefore, the header `WWW-Authenticate` of the negative response `401 Unauthorized` contains the challenge in the parameter `nonce` and the type of hash function in the parameter `algorithm`, which enable Bob to construct his authentication data.

The REGISTRAR concludes the identification of the dialog by filling in the parameter `tag` in the header `To`.

The response is sent to the IP address contained in the header `Via`.

```
SIP/2.0 401 Unauthorized
Via: SIP/2.0/UDP 192.0.2.201;branch=z9hG4bKnashds7
From: Bob <sip:bob@biloxi.com>;tag=a73kszlfl
To: Bob <sip:bob@biloxi.com>;tag=1410948204
Call-ID: 1j9FpLxk3uxtm8tn@biloxi.com
CSeq: 1 REGISTER
WWW-Authenticate: Digest realm="atlanta.com",
qop="auth", nonce="ea9c8e88df84f1cec4341ae6cbe5a359",
opaque="", stale=FALSE, algorithm=MD5
Content-Length: 0
```

Bob's UAS sends a new REGISTER request whose header `Authorization` contains the authentication data in the parameter `response`.

The header `Call-ID` and the parameter `tag` in the header `To` take new values.

The value of the header `Cseq` is incremented by one.

```
REGISTER sip:biloxi.com SIP/2.0
Via: SIP/2.0/UDP 192.0.2.201;branch=z9hG4bKnashd92
```

```
Max-Forwards: 70
From: Bob <sip:bob@biloxi.com>;tag=ja743ks76zlflH
To: Bob <sip:bob@biloxi.com>
Call-ID: 1j9FpLxk3uxtm8tn@biloxi.com
CSeq: 2 REGISTER
Contact: <sip:bob@192.0.2.201>
Authorization: Digest username="bob",
realm="atlanta.com"
nonce="ea9c8e88df84f1cec4341ae6cbe5a359", opaque="",
uri="sip:biloxi.com",
response="dfe56131d1958046689d83306477ecc"
Content-Length: 0
```

The REGISTRAR responds positively if authentication is successful.

The header `Contact` of the response contains the parameter `expires`, which specifies the duration of the registration.

The REGISTRAR completes the identification of the dialog by filling in the parameter `tag` of the header `To`.

```
SIP/2.0 200 OK
Via: SIP/2.0/UDP 192.0.2.201;branch=z9hG4bKnashd92
From: Bob <sip:bob@biloxi.com>;tag=ja743ks76zlflH
To: Bob <sip:bob@biloxi.com>;tag=37GkEhwl6
Call-ID: 1j9FpLxk3uxtm8tn@biloxi.com
CSeq: 2 REGISTER
Contact: <sip:bob@192.0.2.201>;expires=3600
Content-Length: 0
```

4.5.2. *The session*

4.5.2.1. *Establishment of the session*

4.5.2.1.1. Routing of the initial request

Alice's UA uses an SIP application on its terminal (softphone or hardphone) to call Bob. Two PROXY SERVERS, each acting in its own respective domain (atlanta.com and biloxi.com) facilitate the establishment of the session.

The establishment of the session begins with an INVITE request addressed to Bob's SIP URI (`sip:bob@biloxi.com`) (Figure 4.3).

The request contains the headers `Via`, used for the routing of the responses, `Max-Forwards` (value equal to 70), `From` (Alice's SIP URI and parameter `tag`), `To` (Bob's SIP URI), `Call-ID`, `Cseq`, `Contact` (Alice's IP address), `Content-Type` (body of SDP message) and `Content-Length`.

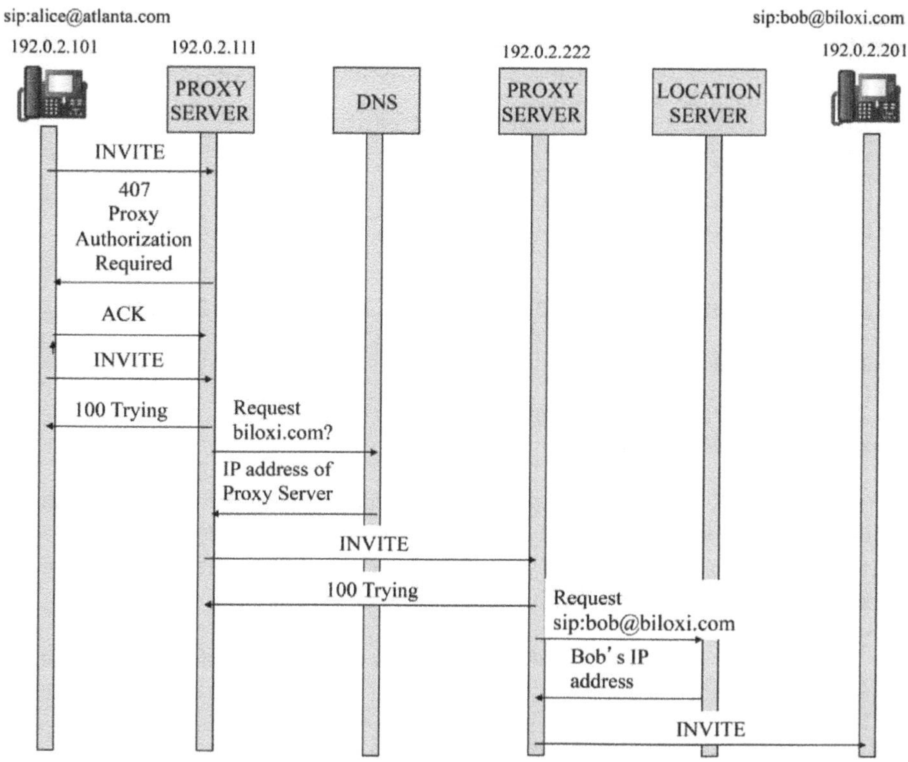

Figure 4.3. *Establishment of the session – routing of the initial request*

As Alice's UA does not know Bob's IP address, it sends the INVITE request to his PROXY SERVER. The IP address of the PROXY SERVER (192.0.2.111) can be configured manually or dynamically (e.g. by way of the DHCP server) in Alice's UA. The IP address of the PROXY SERVER is placed in the header `Route`.

The body of the SDP message contains the characteristics of the session that Alice wishes to establish with Bob:

– IP address: 192.0.2.101;

– port number: 49172;

– type of media: audio;

– type of codec: PCMU.

```
INVITE sip:bob@biloxi.com SIP/2.0
Via: SIP/2.0/UDP 192.0.2.101;branch=z9hG4bK74b43
```

```
Max-Forwards: 70
Route: <sip:192.0.2.111;lr>
From: Alice <sip:alice@atlanta.com>;tag=9fxced76sl
To: Bob <sip:bob@biloxi.com>
Call-ID: 3848276298220188511@atlanta.com
CSeq: 1 INVITE
Contact: <sip:alice@192.0.2.101>
Content-Type: application/sdp
Content-Length: 151

v=0
o=alice 2890844526 2890844526 IN IP4 192.0.2.101
s=-
c=IN IP4 192.0.2.101
t=0 0
m=audio 49172 RTP/AVP 0
a=rtpmap:0 PCMU/8000
```

Alice's PROXY SERVER requires authentication of Alice for the establishment of the session. The header `Proxy-Authenticate` of the response `407 Proxy Authorization Required` contains the challenge in the field `nonce` and specifies the hash algorithm MD5 (Figure 4.3).

The PROXY SERVER supplements the header `To` by adding the parameter `tag`.

```
SIP/2.0 407 Proxy Authorization Required
Via: SIP/2.0/UDP 192.0.2.101;branch=z9hG4bK74b43
From: Alice <sip:alice@atlanta.com>;tag=9fxced76sl
To: Bob <sip:bob@biloxi.com>;tag=3flal12sf
Call-ID: 3848276298220188511@atlanta.com
CSeq: 1 INVITE
Proxy-Authenticate: Digest realm="atlanta.com",
qop="auth", nonce="f84f1cec41e6cbe5aea9c8e88d359",
opaque="", stale=FALSE, algorithm=MD5
Content-Length: 0
```

Alice's UA acknowledges the response received from her PROXY SERVER by sending the ACK message (Figure 4.3).

```
ACK sip:bob@biloxi.com SIP/2.0
Via: SIP/2.0/UDP 192.0.2.101;branch=z9hG4bK74b43
Max-Forwards: 70
From: Alice <sip:alice@atlanta.com>;tag=9fxced76sl
To: Bob <sip:bob@biloxi.com>;tag=3flal12sf
Call-ID: 3848276298220188511@atlanta.com
CSeq: 1 ACK
Content-Length: 0
```

Alice's UA sends back the INVITE request. The header `Proxy-Authorization` contains the parameter `response` whose value is computed on the basis of the value of the field `nonce` received and Alice's secret. The value of the header `Cseq` is incremented by one (Figure 4.3).

It should be noted that the parameter `tag` is absent from the header `To` because it is a new dialog.

```
INVITE sip:bob@biloxi.com SIP/2.0
Via: SIP/2.0/UDP 192.0.2.101;branch=z9hG4bK74bf9
Max-Forwards: 70
Route: <sip: 192.0.2.111;lr>
From: Alice <sip:alice@atlanta.com>;tag=9fxced76sl
To: Bob <sip:bob@biloxi.com>
Call-ID: 3848276298220188511@atlanta.com
CSeq: 2 INVITE
Contact: <sip:alice@192.0.2.101>
Proxy-Authorization: Digest
username="alice",realm="atlanta.com",nonce="wf84f1ceczx4
1ae6cbe5aea9c8e88d359", opaque="",
uri="sip:bob@biloxi.com",
response="42ce3cef44b22f50c6a6071bc8"
Content-Type: application/sdp
Content-Length: 151

v=0
o=alice 2890844526 2890844526 IN IP4 192.0.2.101
s=-
c=IN IP4 192.0.2.101
t=0 0
m=audio 49172 RTP/AVP 0
a=rtpmap:0 PCMU/8000
```

Alice's PROXY SERVER authenticates Alice's UA if the response received in the header `Proxy-Authorization` is identical to the computation carried out locally.

The PROXY SERVER sends a response `100 Trying` to Alice's UA. The response `100 Trying` indicates that the INVITE request has been received and that the PROXY SERVER is processing it to send it to its target. It implicitly confirms the authentication of Alice's UA (Figure 4.3).

```
SIP/2.0 100 Trying
Via: SIP/2.0/UDP 192.0.2.101;branch=z9hG4bK74bf9
From: Alice <sip:alice@atlanta.com>;tag=9fxced76sl
To: Bob <sip:bob@biloxi.com>
```

```
Call-ID: 3848276298220188511@atlanta.com
CSeq: 2 INVITE
Content-Length: 0
```

Alice's PROXY SERVER obtains the IP address of Bob's PROXY SERVER by performing a DNS (Domain Name Service) resolution on the domain name biloxi.com (Figure 4.3).

Before transferring the request, Alice's PROXY SERVER adds new headers `Via` and `Record Route` containing its own address and decreases the value in the field `Max-Forward` by one.

```
INVITE sip:bob@biloxi.com SIP/2.0
Via: SIP/2.0/UDP 192.0.2.111;branch=z9hG4bK2d4790.1
Via: SIP/2.0/UDP 192.0.2.101;branch=z9hG4bK74bf9
Max-Forwards: 69
Record-Route: <sip:192.0.2.111;lr>
From: Alice <sip:alice@atlanta.com>;tag=9fxced76sl
To: Bob <sip:bob@biloxi.com>
Call-ID: 3848276298220188511@atlanta.com
CSeq: 2 INVITE
Contact: <sip:alice@192.0.2.101>
Content-Type: application/sdp
Content-Length: 151

v=0
o=alice 2890844526 2890844526 IN IP4 192.0.2.101
s=-
c=IN IP4 192.0.2.101
t=0 0
m=audio 49172 RTP/AVP 0
a=rtpmap:0 PCMU/8000
```

Bob's PROXY SERVER responds with a `100 Trying` message to Alice's PROXY SERVER to indicate that the INVITE request has been received and that it is working on it to get it to the target (Figure 4.3).

```
SIP/2.0 100 Trying
Via: SIP/2.0/UDP 192.0.2.111;branch=z9hG4bK2d4790.1
Via: SIP/2.0/UDP 192.0.2.101;branch=z9hG4bK74bf9
From: Alice <sip:alice@atlanta.com>;tag=9fxced76sl
To: Bob <sip:bob@biloxi.com>
Call-ID: 3848276298220188511@atlanta.com
CSeq: 2 INVITE
Content-Length: 0
```

Bob's PROXY SERVER consults the LOCATION SERVER which contains Bob's IP address. It replaces the domain name in Bob's ID with his IP address. It adds new headers `Via` and `Record Route`, decreases value of the header `Max-Forwards` by one and sends the INVITE request to Bob's UA (Figure 4.3).

Bob's UA receives the INVITE request and a ringing sound alerts Bob to the arrival of a call.

```
INVITE sip:bob@192.0.2.201 SIP/2.0
Via: SIP/2.0/UDP 192.0.2.222;branch=z9hG4bK721e4.1
Via: SIP/2.0/UDP 192.0.2.111;branch=z9hG4bK2d4790.1
Via: SIP/2.0/UDP 192.0.2.101;branch=z9hG4bK74bf9
Max-Forwards: 68
Record-Route: <sip: 192.0.2.222;lr>,
Record-Route: <sip: 192.0.2.111;lr>
From: Alice <sip:alice@atlanta.com>;tag=9fxced76sl
To: Bob <sip:bob@biloxi.com>
Call-ID: 3848276298220188511@atlanta.com
CSeq: 2 INVITE
Contact: <sip:alice@192.0.2.101>
Content-Type: application/sdp
Content-Length: 151

v=0
o=alice 2890844526 2890844526 IN IP4 192.0.2.101
s=-
c=IN IP4 192.0.2.101
t=0 0
m=audio 49172 RTP/AVP 0
a=rtpmap:0 PCMU/8000
```

4.5.2.1.2. Routing of the response

Bob's UA sends a response `180 Ringing`. Into this response, it copies all the headers `To`, `From`, `Call-ID`, `Cseq`, `Via` and `Record Route` received. It fills the field `Contact` with its own IP address. It sends the response to the address contained in the first header `Via` (Figure 4.4).

Bob's UA adds a parameter `tag` to the header `To`, which enables us to complete the identifier of the dialog initialized with the header `Call-ID` and the parameter `tag` in the header `From`.

```
SIP/2.0 180 Ringing
Via: SIP/2.0/UDP 192.0.2.222;branch=z9hG4bK721e4.1
Via: SIP/2.0/UDP 192.0.2.111;branch=z9hG4bK2d4790.1
Via: SIP/2.0/UDP 192.0.2.101;branch=z9hG4bK74bf9
Record-Route: <sip: 192.0.2.222;lr>,
Record-Route: <sip: 192.0.2.111;lr>
From: Alice <sip:alice@atlanta.com>;tag=9fxced76sl
To: Bob <sip:bob@biloxi.com>;tag=314159
Call-ID: 3848276298220188511@atlanta.com
```

```
Contact: <sip:bob@192.0.2.201>
CSeq: 2 INVITE
Content-Length: 0
```

Bob's PROXY SERVER deletes the `Via` header containing its own address and uses the next `Via` to determine the destination of the response (Figure 4.4).

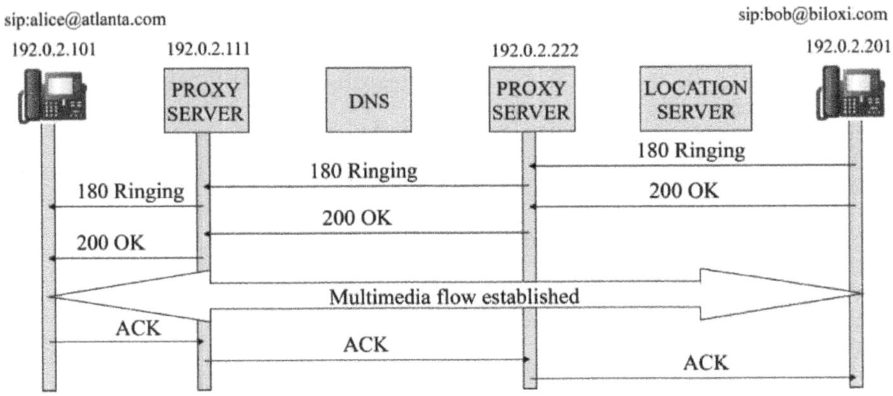

Figure 4.4. *Establishment of the session – routing of the responses*

```
SIP/2.0 180 Ringing
Via: SIP/2.0/UDP 192.0.2.111;branch=z9hG4bK2d4790.1
Via: SIP/2.0/UDP 192.0.2.101;branch=z9hG4bK74bf9
Record-Route: <sip: 192.0.2.222;lr>,
Record-Route: <sip: 192.0.2.111;lr>
From: Alice <sip:alice@atlanta.com>;tag=9fxced76sl
To: Bob <sip:bob@biloxi.com>;tag=314159
Call-ID: 3848276298220188511@atlanta.com
Contact: <sip:bob@192.0.2.201>
CSeq: 2 INVITE
Content-Length: 0
```

Alice's PROXY SERVER deletes the `Via` header containing its own address and uses the next `Via` to determine the destination of the response (Figure 4.4).

When Alice's UA receives the response `180 Ringing`, a ring-back tone is presented to Alice.

```
SIP/2.0 180 Ringing
Via: SIP/2.0/UDP 192.0.2.101;branch=z9hG4bK74bf9
Record-Route: <sip: 192.0.2.222;lr>,
Record-Route: <sip: 192.0.2.111;lr>
From: Alice <sip:alice@atlanta.com>;tag=9fxced76sl
To: Bob <sip:bob@biloxi.com>;tag=314159
```

```
Call-ID: 3848276298220188511@atlanta.com
Contact: <sip:bob@192.0.2.201>
CSeq: 2 INVITE
Content-Length: 0
```

When Bob picks up his telephone handset, his UA sends the definitive response `200 OK` which contains all the headers received in the INVITE request (Figure 4.4).

Bob's UA adds the body of the SDP message containing the description of the session which Bob is ready to establish with Alice:

– IP address: 192.0.2.201;

– port number: 3456;

– type of media: audio;

– type of codec: PCMU.

```
SIP/2.0 200 OK
Via: SIP/2.0/UDP 192.0.2.222;branch=z9hG4bK721e4.1
Via: SIP/2.0/UDP 192.0.2.111;branch=z9hG4bK2d4790.1
Via: SIP/2.0/UDP 192.0.2.101;branch=z9hG4bK74bf9
Record-Route: <sip: 192.0.2.222;lr>,
Record-Route: <sip: 192.0.2.111;lr>
From: Alice <sip:alice@atlanta.com>;tag=9fxced76sl
To: Bob <sip:bob@biloxi.com>;tag=314159
Call-ID: 3848276298220188511@atlanta.com
CSeq: 2 INVITE
Contact: <sip:bob@192.0.2.201>
Content-Type: application/sdp
Content-Length: 147

v=0
o=bob 2890844527 2890844527 IN IP4 192.0.2.201
s=-
c=IN IP4 192.0.2.201
t=0 0
m=audio 3456 RTP/AVP 0
a=rtpmap:0 PCMU/8000
```

The response is relayed by Bob's PROXY SERVER to Alice's PROXY SERVER (Figure 4.4).

```
SIP/2.0 200 OK
Via: SIP/2.0/UDP 192.0.2.111;branch=z9hG4bK2d4790.1
Via: SIP/2.0/UDP 192.0.2.101;branch=z9hG4bK74bf9
Record-Route: <sip: 192.0.2.222;lr>,
Record-Route: <sip: 192.0.2.111;lr>
From: Alice <sip:alice@atlanta.com>;tag=9fxced76sl
```

```
To: Bob <sip:bob@biloxi.com>;tag=314159
Call-ID: 3848276298220188511@atlanta.com
CSeq: 2 INVITE
Contact: <sip:bob@192.0.2.201>
Content-Type: application/sdp
Content-Length: 147

v=0
o=bob 2890844527 2890844527 IN IP4 192.0.2.201
s=-
c=IN IP4 192.0.2.201
t=0 0
m=audio 3456 RTP/AVP 0
a=rtpmap:0 PCMU/8000
```

The response `200 OK` is relayed by Alice's PROXY SERVER to Alice's UA, which stops the ring-back tone and triggers the establishment of the media flow (Figure 4.4).

```
SIP/2.0 200 OK
Via: SIP/2.0/UDP 192.0.2.101;branch=z9hG4bK74bf9
Record-Route: <sip: 192.0.2.222;lr>,
Record-Route: <sip: 192.0.2.111;lr>
From: Alice <sip:alice@atlanta.com>;tag=9fxced76sl
To: Bob <sip:bob@biloxi.com>;tag=314159
Call-ID: 3848276298220188511@atlanta.com
CSeq: 2 INVITE
Contact: <sip:bob@192.0.2.201>
Content-Type: application/sdp
Content-Length: 147

v=0
o=bob 2890844527 2890844527 IN IP4 192.0.2.201
s=-
c=IN IP4 192.0.2.201
t=0 0
m=audio 3456 RTP/AVP 0
a=rtpmap:0 PCMU/8000
```

4.5.2.1.3. Routing of the subsequent request

Alice's UA sends an acknowledgement ACK to Bob's UA to confirm receipt of the response `200 OK` (Figure 4.4). The request includes the headers `Route` which contain the list of PROXY SERVERS that the request must pass through. These headers are filled in on the basis of the content of the headers `Record Route` received in the responses.

```
ACK sip:bob@192.0.2.201 SIP/2.0
Via: SIP/2.0/UDP 192.0.2.101;branch=z9hG4bK74b76
```

```
Max-Forwards: 70
Route: <sip:192.0.2.111;lr>,
Route: <sip:192.0.2.222;lr>
From: Alice <sip:alice@atlanta.com>;tag=9fxced76sl
To: Bob <sip:bob@biloxi.com>;tag=314159
Call-ID: 3848276298220188511@atlanta.com
CSeq: 2 ACK
Content-Length: 0
```

The ACK request is relayed by Alice's PROXY SERVER to Bob's PROXY SERVER (Figure 4.4).

```
ACK sip:bob@192.0.2.201 SIP/2.0
Via: SIP/2.0/UDP 192.0.2.111;branch=z9hG4bK2d4790.1
Via: SIP/2.0/UDP 192.0.2.101;branch=z9hG4bK74b76
Max-Forwards: 69
Route: <sip:192.0.2.222;lr>
From: Alice <sip:alice@atlanta.com>;tag=9fxced76sl
To: Bob <sip:bob@biloxi.com>;tag=314159
Call-ID: 3848276298220188511@atlanta.com
CSeq: 2 ACK
Content-Length: 0
```

The ACK request is relayed by Bob's PROXY SERVER to Bob's UA (Figure 4.4).

```
ACK sip:bob@192.0.2.201 SIP/2.0
Via: SIP/2.0/UDP 192.0.2.222;branch=z9hG4bK721e4.1
Via: SIP/2.0/UDP 192.0.2.111;branch=z9hG4bK2d4790.1
Via: SIP/2.0/UDP 192.0.2.101;branch=z9hG4bK74b76
Max-Forwards: 68
From: Alice <sip:alice@atlanta.com>;tag=9fxced76sl
To: Bob <sip:bob@biloxi.com>;tag=314159
Call-ID: 3848276298220188511@atlanta.com
CSeq: 2 ACK
Content-Length: 0
```

During the session, Alice or Bob may decide to modify the characteristics of the media, e.g. to switch from a telephone session to a videophone session, sending a new INVITE request (reINVITE) containing the description of the new media. This new request contains the same identifiers as the dialog so that the other party knows that it is indeed a modification of the existing session.

The other party sends a response `200 OK` to accept the change. The asker responds with an acknowledgement message ACK. If the other party does not accept the change, it sends back a definitive negative response `488 Not acceptable`, and the asker also sends an acknowledgement. However, the failure of the INVITE request does not terminate the call in progress, and the session continues to use the characteristics previously negotiated.

4.5.2.2. *Release of the session*

4.5.2.2.1. Case 1 – the session is established

At the end of the communication, if Bob hangs up first, his UA sends a BYE message to his PROXY SERVER (Figure 4.5).

```
BYE sip:alice@192.0.2.101 SIP/2.0
Via: SIP/2.0/UDP 192.0.2.201;branch= z9hG4bK74b03
Max-Forwards: 70
Route: <sip:192.0.2.222;lr>,
Route: <sip:192.0.2.111;lr>
From: Bob <sip:bob@biloxi.com>;tag=314159
To: Alice <sip:alice@atlanta.com>;tag=9fxced76sl
Call-ID: 3848276298220188511@atlanta.com
CSeq: 1 BYE
Content-Length: 0
```

The BYE request is relayed by Bob's PROXY SERVER to Alice's PROXY SERVER (Figure 4.5).

```
BYE sip:alice@192.0.2.101 SIP/2.0
Via: SIP/2.0/UDP 192.0.2.222;branch= z9hG4bK74b21
Via: SIP/2.0/UDP 192.0.2.201;branch= z9hG4bK74b03
Max-Forwards: 69
Route: <sip:192.0.2.101;lr>
From: Bob <sip:bob@biloxi.com>;tag=314159
To: Alice <sip:alice@atlanta.com>;tag=9fxced76sl
Call-ID: 3848276298220188511@atlanta.com
CSeq: 1 BYE
Content-Length: 0
```

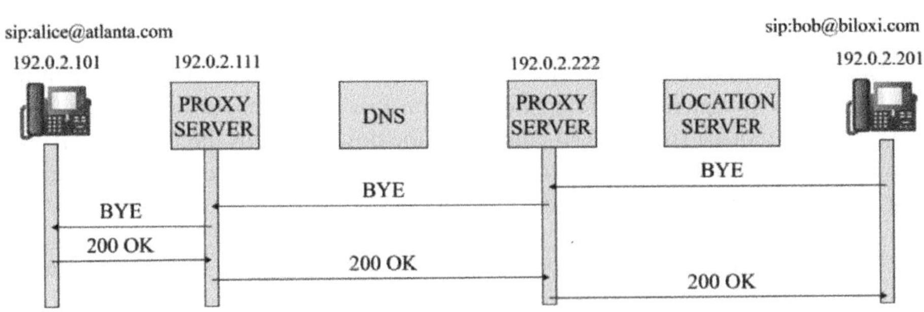

Figure 4.5. *Release of the session – the session is established*

The BYE request is relayed by Alice's PROXY SERVER to Alice's UA (Figure 4.5).

```
BYE sip:alice@192.0.2.101 SIP/2.0
Via: SIP/2.0/UDP 192.0.2.111;branch= z9hG4bK230f2.1
Via: SIP/2.0/UDP 192.0.2.222;branch= z9hG4bK74b21
Via: SIP/2.0/UDP 192.0.2.201;branch= z9hG4bK74b03
Max-Forwards: 68
From: Bob <sip:bob@biloxi.com>;tag=314159
To: Alice <sip:alice@atlanta.com>;tag=9fxced76sl
Call-ID: 3848276298220188511@atlanta.com
CSeq: 1 BYE
Content-Length: 0
```

Alice's UA confirms receipt of the BYE request with a response `200 OK`, which terminates the dialog (Figure 4.5).

```
SIP/2.0 200 OK
Via: SIP/2.0/UDP 192.0.2.111;branch= z9hG4bK230f2.1
Via: SIP/2.0/UDP 192.0.2.222;branch= z9hG4bK74b21
Via: SIP/2.0/UDP 192.0.2.201;branch= z9hG4bK74b03
From: Bob <sip:bob@biloxi.com>;tag=314159
To: Alice <sip:alice@atlanta.com>;tag=9fxced76sl
Call-ID: 3848276298220188511@atlanta.com
CSeq: 1 BYE
Content-Length: 0
```

The response `200 OK` is relayed by Alice's PROXY SERVER to Bob's PROXY SERVER (Figure 4.5).

```
SIP/2.0 200 OK
Via: SIP/2.0/UDP 192.0.2.222;branch= z9hG4bK74b21
Via: SIP/2.0/UDP 192.0.2.201;branch= z9hG4bK74b03
From: Bob <sip:bob@biloxi.com>;tag=314159
To: Alice <sip:alice@atlanta.com>;tag=9fxced76sl
Call-ID: 3848276298220188511@atlanta.com
CSeq: 1 BYE
Content-Length: 0
```

The response `200 OK` is relayed by Bob's PROXY SERVER to Bob's UA (Figure 4.5).

```
SIP/2.0 200 OK
Via: SIP/2.0/UDP 192.0.2.201;branch= z9hG4bK74b03
From: Bob <sip:bob@biloxi.com>;tag=314159
To: Alice <sip:alice@atlanta.com>;tag=9fxced76sl
Call-ID: 3848276298220188511@atlanta.com
CSeq: 1 BYE
Content-Length: 0
```

4.5.2.2.2. Case 2 – the callee does not answer

If Bob does not pick up his handset, Alice can hang up and terminate a session which has not been successfully established. Alice's UA triggers the termination of the dialog initialized by the INVITE request by sending the CANCEL request (Figure 4.6).

```
CANCEL sip:bob@192.0.2.201 SIP/2.0
Via: SIP/2.0/UDP 192.0.2.101;branch= z9hG4bK74b44
Max-Forwards: 70
From: Alice <sip:alice@atlanta.com>;tag=9fxced76sl
To: Bob <sip:bob@biloxi.com>;tag=314159
Route: <sip:192.0.2.111;lr>,
Route: <sip:192.0.2.222;lr>
Call-ID: 3848276298220188511@atlanta.com
CSeq: 1 CANCEL
Content-Length: 0
```

Alice's PROXY SERVER transfers the CANCEL request to Bob's PROXY SERVER (Figure 4.6).

```
CANCEL sip:bob@192.0.2.201 SIP/2.0
Via: SIP/2.0/UDP 192.0.2.111;branch= z9hG4bK31972.1
Via: SIP/2.0/UDP 192.0.2.101;branch= z9hG4bK74b44
Max-Forwards: 69
From: Alice <sip:alice@atlanta.com>;tag=9fxced76sl
To: Bob <sip:bob@biloxi.com>;tag=314159
Route: <sip:192.0.2.222;lr>
Call-ID: 3848276298220188511@atlanta.com
CSeq: 1 CANCEL
Content-Length: 0
```

Bob's PROXY SERVER transfers the CANCEL request to Bob's UA (Figure 4.6).

```
CANCEL sip:bob@192.0.2.201 SIP/2.0
Via: SIP/2.0/UDP 192.0.2.222;branch= z9hG4bK465b6d
Via: SIP/2.0/UDP 192.0.2.111;branch= z9hG4bK31972.1
Via: SIP/2.0/UDP 192.0.2.101;branch= z9hG4bK74b44
Max-Forwards: 68
From: Alice <sip:alice@atlanta.com>;tag=9fxced76sl
To: Bob <sip:bob@biloxi.com>;tag=314159
Call-ID: 3848276298220188511@atlanta.com
CSeq: 1 CANCEL
Content-Length: 0
```

Bob's UA responds to the CANCEL request to terminate the transaction (Figure 4.6).

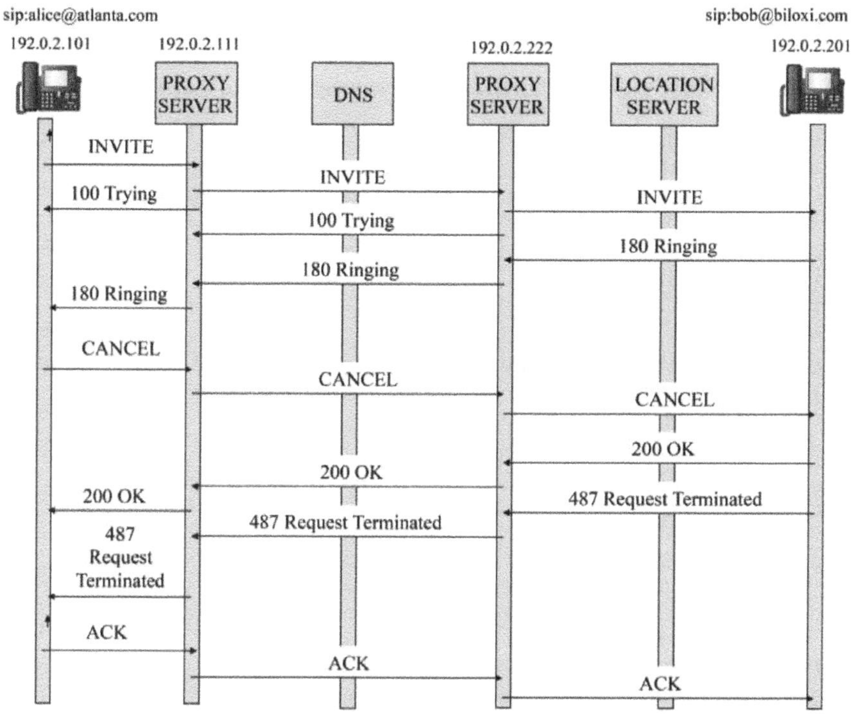

Figure 4.6. *Release of the session – the callee does not answer*

```
SIP/2.0 200 OK
Via: SIP/2.0/UDP 192.0.2.222;branch= z9hG4bK465b6d
Via: SIP/2.0/UDP 192.0.2.111;branch= z9hG4bK31972.1
Via: SIP/2.0/UDP 192.0.2.101;branch= z9hG4bK74b44
From: Alice <sip:alice@atlanta.com>;tag=9fxced76sl
To: Bob <sip:bob@biloxi.com>;tag=314159
Call-ID: 3848276298220188511@atlanta.com
CSeq: 1 CANCEL
Content-Length: 0
```

Bob's PROXY SERVER transfers the response to Alice's PROXY SERVER (Figure 4.6).

```
SIP/2.0 200 OK
Via: SIP/2.0/UDP 192.0.2.111;branch= z9hG4bK31972.1
Via: SIP/2.0/UDP 192.0.2.101;branch= z9hG4bK74b44
From: Alice <sip:alice@atlanta.com>;tag=9fxced76sl
To: Bob <sip:bob@biloxi.com>;tag=314159
Call-ID: 3848276298220188511@atlanta.com
```

```
CSeq: 1 CANCEL
Content-Length: 0
```

Alice's PROXY SERVER transfers the response to Alice's UA (Figure 4.6).

```
SIP/2.0 200 OK
Via: SIP/2.0/UDP 192.0.2.101;branch= z9hG4bK74b44
From: Alice <sip:alice@atlanta.com>;tag=9fxced76sl
To: Bob <sip:bob@biloxi.com>;tag=314159
Call-ID: 3848276298220188511@atlanta.com
CSeq: 1 CANCEL
Content-Length: 0
```

Bob's UA terminates the dialog initialized by the INVITE request, which stops the ringing (Figure 4.6).

```
SIP/2.0 487 Request Terminated
Via: SIP/2.0/UDP 192.0.2.222;branch=z9hG4bK721e4.1
Via: SIP/2.0/UDP 192.0.2.111;branch=z9hG4bK2d4790.1
Via: SIP/2.0/UDP 192.0.2.101;branch=z9hG4bK74bf9
From: Alice <sip:alice@atlanta.com>;tag=9fxced76sl
To: Bob <sip:bob@biloxi.com>;tag=314159
From: Alice <sip:alice@atlanta.com>;tag=9fxced76sl
To: Bob <sip:bob@biloxi.com>;tag=314159
Call-ID: 3848276298220188511@atlanta.com
CSeq: 1 INVITE
Content-Length: 0
```

Bob's PROXY SERVER transfers the response to Alice's PROXY SERVER (Figure 4.6).

```
SIP/2.0 487 Request Terminated
Via: SIP/2.0/UDP 192.0.2.111;branch=z9hG4bK2d4790.1
Via: SIP/2.0/UDP 192.0.2.101;branch=z9hG4bK74bf9
From: Alice <sip:alice@atlanta.com>;tag=9fxced76sl
To: Bob <sip:bob@biloxi.com>;tag=314159
Call-ID: 3848276298220188511@atlanta.com
CSeq: 1 INVITE
Content-Length: 0
```

Alice's PROXY SERVER transfers the response to Alice's UA, which stops the ringback tone (Figure 4.6).

```
SIP/2.0 487 Request Terminated
Via: SIP/2.0/UDP 192.0.2.101;branch=z9hG4bK74bf9
From: Alice <sip:alice@atlanta.com>;tag=9fxced76sl
To: Bob <sip:bob@biloxi.com>;tag=314159
Call-ID: 3848276298220188511@atlanta.com
CSeq: 1 INVITE
```

Alice's UA sends an acknowledgement message ACK to Bob's UA to confirm receipt of the response `487 Request Terminated` (Figure 4.6).

4.5.2.2.3. Case 3 – the callee is busy

On receiving the INVITE request, Bob's UA transmits the definitive negative response `486 Busy Here`, showing that Bob is in another communication and cannot take the call (Figure 4.7).

```
SIP/2.0 486 Busy Here
Via: SIP/2.0/UDP 192.0.2.222;branch=z9hG4bK721e4.1
Via: SIP/2.0/UDP 192.0.2.111;branch=z9hG4bK2d4790.1
Via: SIP/2.0/UDP 192.0.2.101;branch=z9hG4bK74bf9
From: Alice <sip:alice@atlanta.com>;tag=9fxced76sl
To: Bob <sip:bob@biloxi.com>;tag=314159
Call-ID: 3848276298220188511@atlanta.com
CSeq: 1 INVITE
Content-Length: 0
```

Bob's PROXY SERVER transfers the response to Alice's PROXY SERVER (Figure 4.7).

```
SIP/2.0 486 Busy Here
Via: SIP/2.0/UDP 192.0.2.111;branch=z9hG4bK2d4790.1
Via: SIP/2.0/UDP 192.0.2.101;branch=z9hG4bK74bf9
From: Alice <sip:alice@atlanta.com>;tag=9fxced76sl
To: Bob <sip:bob@biloxi.com>;tag=314159
Call-ID: 3848276298220188511@atlanta.com
CSeq: 1 INVITE
Content-Length: 0
```

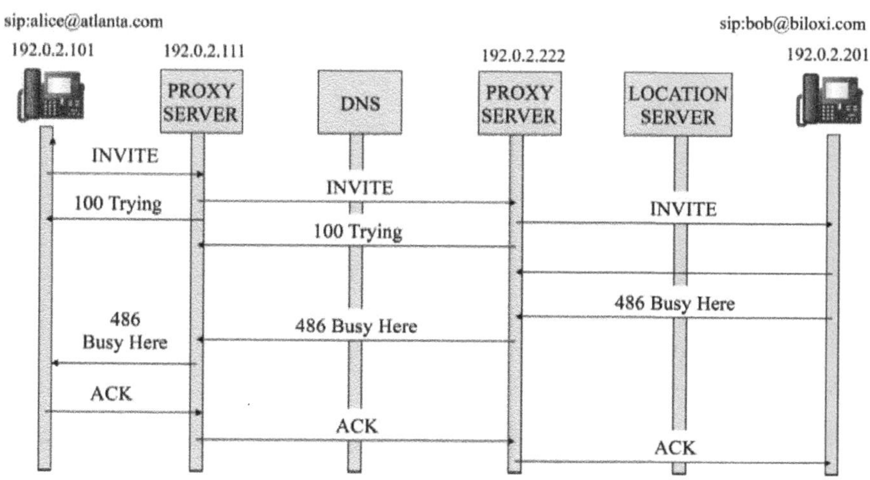

Figure 4.7. *Release of the session – the callee is busy*

Alice's PROXY SERVER transfers the response to Alice's UA. Receipt of the message triggers the busy tone (Figure 4.7).

```
SIP/2.0 486 Busy Here
Via: SIP/2.0/UDP 192.0.2.101;branch=z9hG4bK74bf9
From: Alice <sip:alice@atlanta.com>;tag=9fxced76sl
To: Bob <sip:bob@biloxi.com>;tag=314159
Call-ID: 3848276298220188511@atlanta.com
CSeq: 1 INVITE
Content-Length: 0
```

Alice's UA sends an acknowledgement ACK to Bob's UA to confirm receipt of the response `486 Busy Here` (Figure 4.7).

Chapter 5

The IMS Network

5.1. Architecture of IMS

The IMS network (IP Multimedia Subsystem) provides the telephone service when the mobile network uses PS (Packet Service) mode.

The IMS network connects to the following entities (Figure 5.1):

– PGW (PDN (Packet Data Network) Gateway) of the EPC (Evolved Packet Core) in the EPS (Evolved Packet System) network. The EPS network constructs two bearers: one for transport of the signaling exchanged with the mobile and the other for transport of the media (voice, video or data);

– PCRF (Policy Charging and Rules Function) for control of the media;

– HSS (Home Subscriber Server) for access to the profile and security data of the mobile;

– networks PSTN (Public Switched Telephone Network) and PLMN (Public Land Mobile Network) for the interconnection.

The IMS network includes the following functions (Figure 5.2):

– Call Session Control Function (CSCF), involving P-CSCF (Proxy-CSCF), S-CSCF (Serving-CSCF), I-CSCF (Interrogating-CSCF) and E-CSCF (Emergency-CSCF);

– Application Servers (AS);

– Multimedia Resource Function (MRF), involving MRFC (MRF Controller) and MFRP (MRF Processor);

– interconnection with the PSTN or PLMN, involving BGCF (Breakout Gateway Control Function), MGCF (Media Gateway Control Function), MGW (Media Gateway) and SGW (Signaling Gateway);

– offline charging for post-paid and online charging for pre-paid.

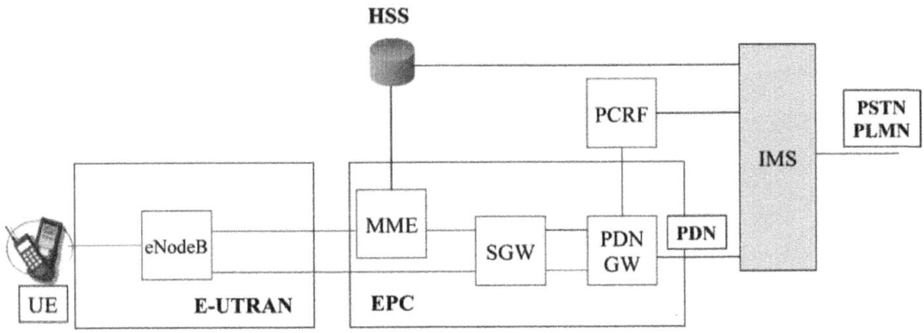

Figure 5.1. *Positioning of the IMS network*

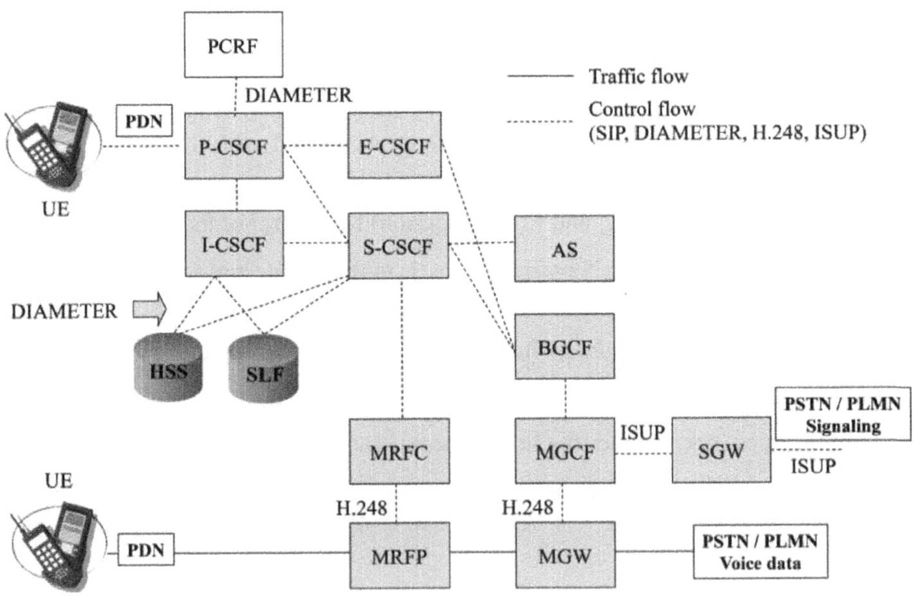

Figure 5.2. *Entities in the IMS network*

5.1.1. *Session control*

5.1.1.1. *P-CSCF*

The P-CSCF is the first point of contact for the mobile UE (User Equipment) in the IMS network. It performs the function of a PROXY SERVER. It receives the requests from the UE or from the S-CSCF and transfers them respectively to the S/I-CSCF or to the UE.

The P-CSCF may also act as a UA (User Agent) in abnormal operating conditions, when it has to terminate or generate SIP (Session Information Protocol) transactions.

The tasks performed by the P-CSCF are:

– transfering the SIP request REGISTER to the I-CSCF determined on the basis of the domain name provided by the UE. Into this message, it adds a header `Path` containing its IP address. This address is preserved by the S-CSCF;

– transfering the SIP request INVITE received from the S-CSCF (or respectively from the UE) to the UE (or respectively to the S-CSCF). To carry out the transfer, the P-CSCF has to find the IP addresses of the UE (or respectively of the S-CSCF). The SIP request INVITE received from the S-CSCF contains the IP address of the UE instead of the URI (Uniform Resource Identifier). The SIP request INVITE received from the UE contains the IP address of the S-CSCF in the header `Route`;

– detecting emergency calls and transfering them to the E-CSCF;

– generating the data necessary for the generation of charging tickets;

– establishing an IPSec (IP Security) association with the UE at its registration;

– using the DIAMETER messages exchanged with the PCRF to control the type of resources required by the UE on the basis of the capacities authorized by the EPS network;

– checking resource availability in the EPS network.

5.1.1.2. *I-CSCF*

The I-CSCF is the point of contact within the IMS network for some transactions coming from the P-CSCF or S-CSCF. It performs the function of a PROXY SERVER.

The tasks performed by the I-CSCF are:

– upon receipt of the first SIP REGISTER request, assigning an S-CSCF to the UE and transfers the request to the S-CSCF chosen. To fulfill this function, an exchange of DIAMETER messages with the HSS entity is necessary;

– upon receipt of the second SIP REGISTER request and the first SIP INVITE request, for an incoming call, querying the HSS for the IP address of the S-CSCF attributed to the UE and transfers the request to that S-CSCF;

– generating the data necessary for the generation of charging tickets.

5.1.1.3. *S-CSCF*

The S-CSCF provides the UE with session control services. It performs different roles depending on the type of request received:

– that of a REGISTRAR for the registration of the UE;

– that of a LOCATION SERVER for the storage of the correspondence between the IP address and the URI of the UE;

– that of a PROXY SERVER for the establishment of a session;

– that of a UA, in abnormal operating conditions, when it has to terminate or generate SIP transactions.

The tasks performed in the REGISTRAR function are:

– On receiving the first REGISTER request, it contacts the HSS to recover the UE authentication data. To fulfill this function, an exchange of DIAMETER messages with the HSS is required. It responds with a message `401 Unauthorized` containing the parameters used for authentication.

– On receiving the second REGISTER request, it authenticates the UE and recovers its profile from the HSS. It responds with a message `200 OK` containing a header `Service Route` containing its IP address which the UA keeps in its memory.

The tasks performed in the PROXY SERVER function are:

– For an outgoing call, on receipt of the first SIP INVITE request from the P-CSCF, it performs a check on the service required on the basis of the profile recovered during registration. It transfers the request either to the AS or to the I-CSCF belonging to the IMS network of the required UE, or to the BGCF if the required UE is not in an IMS network. The IP address of the AS is contained in the profile of the UE recovered during registration.

– For an incoming call, on receipt of the first SIP INVITE request from the I-CSCF, it performs a check on the service required. It transfers the request either to the AS or to the P-CSCF. In the latter case, it replaces the URI of the request with the IP address of the UE. The IP address of the P-CSCF is recovered on the basis of the header `Path`, during the registration of the UE.

– It generates the data necessary to generate charging tickets.

5.1.1.4. *E-CSCF*

The E-CSCF handles emergency calls transmitted by the P-CSCF and routes the request to the emergency center nearest to the UE. The emergency center can be linked to a fixed or mobile network or to another IMS network.

The tasks performed by the E-CSCF are:

– On receiving the SIP request INVITE, it contacts the LRF (Location Retrieval Function) to obtain the location of the UE or to validate if it is included in the request.

– On the basis of information also provided by the LRF, it transfers the request to the nearest emergency center.

5.1.2. ***Application servers***

The AS provides added-value services to the IMS network. For instance, it hosts and executes the supplementary telephone services.[1] It may affect the SIP session depending on the service required. It may be located within the home network or be provided by a third party.

The S-CSCF has to decide whether an AS is necessary to receive the information relating to an SIP request in order to ensure handling of the appropriate service. The decision is based on the information received from the HSS during user registration.

The AS may play various roles in the processing of an SIP message:

– that of a PROXY SERVER. In this mode, the SIP request from the S-CSCF is sent to the AS. The AS can add, remove or modify headers in the SIP message;

– that of a UAS (UA Server) or a REDIRECT SERVER. In this mode, the response of the AS to the SIP request from the S-CSCF is `2xx`, `4xx`, `5xx`, `6xx` (UAS) or `3xx` (REDIRECT SERVER);

---

1 Telephone services are described in Chapter 6.

– that of a UAC (UA Client). In this mode, the AS generates the SIP request and transmits it to the S-CSCF;

– that of a B2BUA. In this mode, the AS receiving an SIP request from the S-CSCF terminates the dialog and generates a new request.

5.1.3. *Databases*

The HSS is a database storing the data specific to each user. The main data stored include the IDs of the users, the access parameters and the rules of invocation of the application servers by the S-CSCF.

The user IDs include public and private IDs. A private ID is an ID assigned by the operator of the IMS network. It is used for registration. A public ID is the ID that other users can use to establish a session.

The access parameters are used for the authentication of the user during registration. The service triggering information is used by the S-CSCF to transfer an SIP request to an AS.

The SLF (Subscription Locator Functional) enables the CSCFs to find the address of the HSS assigned to a UE, when several HSS are deployed.

5.1.4. *Interconnection*

The BGCF processes the INVITE requests sent by the S-CSCF in case the session cannot be forwarded to an IMS network. This relates to calls to the users connected to the PSTN or PLMN networks.

The BGCF determines the next hop for the routing of the SIP message. It has to choose the MGCF that is responsible for internetworking with the PSTN or PLMN. If the interconnection entity is in a third-party network, it transmits the SIP message to another BGCF in that third-party network.

The MGCF takes care of the establishment, maintenance and release of the connections in an MGW. A connection represents an association between an input end-point (the interface with the third-party network PSTN and PLMN) and an output end-point (the interface with the IP network) and vice versa. The MGCF uses the H.248 or MeGaCo (Media Gateway Controller) protocol to control multiple MGWs.

The SGW performs the conversion of the transport protocols relating to the signaling exchanged between the MGCF (SIGTRAN (Signaling Transport over IP) model) and the PSTN and PLMN (SS7 (Signaling System 7) model) (Figure 5.3).

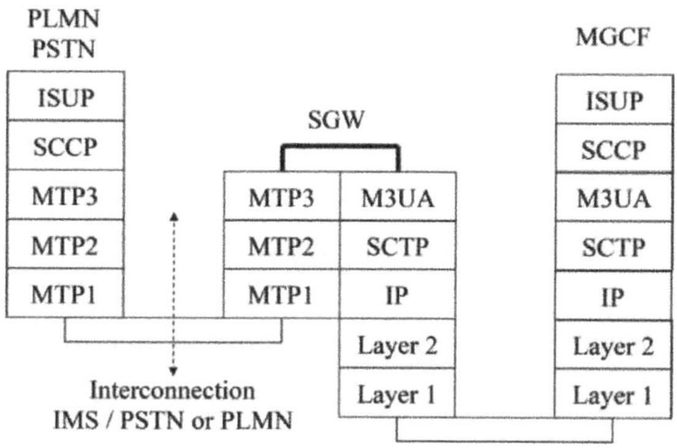

Figure 5.3. *Transport of signaling data*

The MGW performs the conversion of protocols relating to the multimedia flows between the two end-points (Figure 5.4). It contains the treatments carried out on the media flows, such as transcoding (modification of the type of codec between the two end-points), echo cancelling and transmission of tones and announcements.

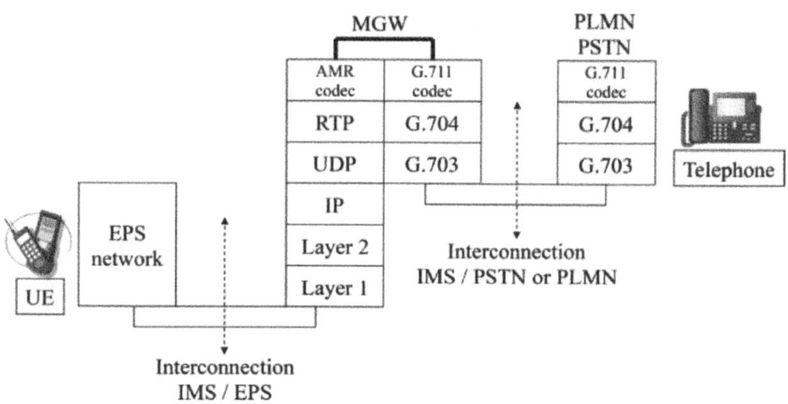

Figure 5.4. *Transport of voice data*

5.1.5. *Media processing*

Media processing is done by the MRF function, divided into two entities: the MRFC and MRFP.

The tasks performed by the MRFC are as follows:

– it controls the media resources of the MRFP;

– it interprets information from the S-CSCF and controls the MRFP on the basis of this interpretation;

– it generates the data necessary for the generation of charging tickets.

The tasks performed by the MRFP are as follows:

– it generates media flows under the control of the MRFC (such as telephone announcements);

– it combines media flows to provide a conferencing service;

– it can also perform particular treatments of the media flows, such as the transcoding of the audio signal.

5.1.6. *Charging*

To assist the charging mechanisms, the IMS network monitors resource use in real time to detect events relevant to charging.

In the case of offline charging, the resource use is notified after the session. This mode of charging corresponds to post-paid service.

In the case of online charging, the user's account is consulted before the granting of permission to use the network resource. That account is decreased during the communication. When it reaches zero, the communication is cut off. This mode of charging corresponds to pre-paid service.

5.1.6.1. *Offline charging*

The CTF (Charging Trigger Function) generates charging events based on the observation of the use of network resources. It is integrated in all the entities of the IMS network (Figure 5.5).

The CDF (Charging Data Function) receives the charging data from the CTF. It then uses these data to generate charging tickets – CDRs (Charging Data Records). DIAMETER is the message exchange protocol used between the two functions (Figure 5.5).

The CDRs produced by the CDF are kept in the CGF (Charging Gateway Function) – a database which acts as a gateway with the billing system (Figure 5.5).

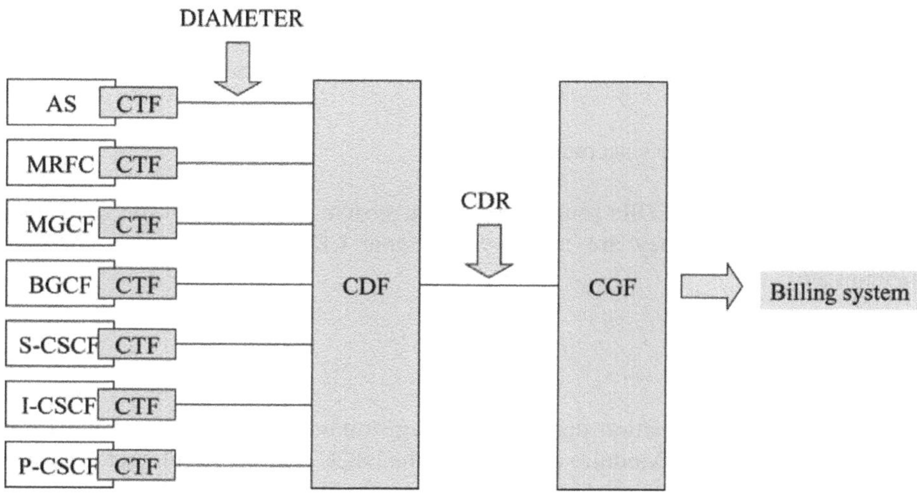

Figure 5.5. *Offline charging*

5.1.6.2. *Online charging*

The S-CSCF does not trigger any charging events and does not include CTF. Online charging is transparent for this function, appearing as a service logic controlled by an IMS-GWF (IMS Gateway Function) AS (Figure 5.6).

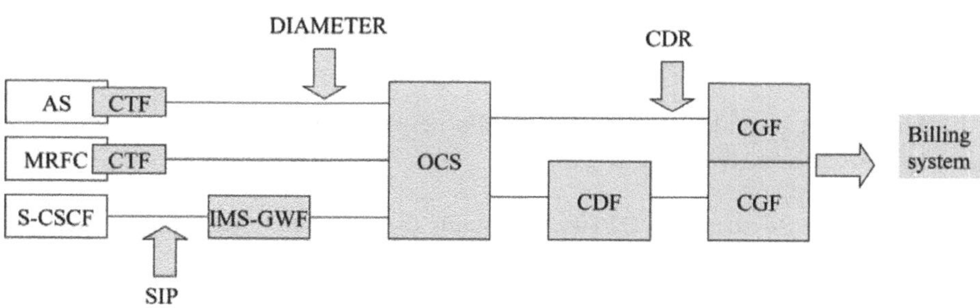

Figure 5.6. *Online charging*

The OCS (Online Charging System) comprises several distinct modules (Figure 5.6):

– charging on the basis of the sessions established by the users (e.g. voice calls);

– charging on the basis of events in conjunction with application servers;

– valorization of the use of network resources in order to calculate the amount of charging;

– balance of the user's account.

The generation of CDRs sent to the billing system is optional. If the OCS does not produce CDRs, they are used by the same CDF as with offline charging (Figure 5.6).

5.2. Registration

The relevant information necessary for registration is obtained from the ISIM (IMS Services Identity Module) contained in the UICC (Universal Integrated Circuit Card) of the UE. The registration takes place in two phases so as to authenticate the mobile. If the authentication of the mobile by the EPS network can be recovered by the IMS network, in this case registration takes place in a single phase.

5.2.1. *First phase of registration*

The first phase of registration includes the steps shown in Figure 5.7.

Alice's UA sends her P-CSCF an initial REGISTER request containing her private ID `alice_private@atlanta.com` in the parameter `username` in the header `Authorization`.

Alice's UA provides the type of mobile network and the ID of the cell in the header `P-Access-Network-Info`.

The header `Security-Client` contains the parameters defining the security association IPSec established with the P-CSCF.

The headers `Require` and `Proxy-Require` contain the value `sec-agree`, indicating that the security mechanism based on IPSec is required. The header `Proxy-Require` is addressed to the P-CSCF. The header `Require` is addressed to the S-CSCF in case the request is transmitted directly to it.

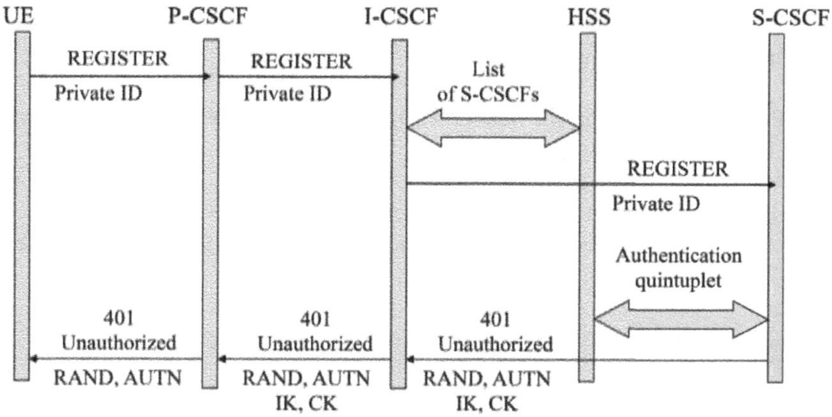

Figure 5.7. *First phase of registration*

```
REGISTER sip:registrar.atlanta.com SIP/2.0
Via: SIP/2.0/UDP 192.0.2.101;branch=z9hG4bKnashds7
Max-Forwards: 70
P-Access-Network-Info: 3GPP-EUTRAN-FDD; eutran-cell-id-
3gpp=234151D0FCE11
From: <sip:alice@atlanta.com>;tag=4fa3
To: <sip:alice@atlanta.com >
Contact: <sip:alice@192.0.2.101>;expires=600000
Call-ID: apb03a0s09dkjdfglkj49111
Authorization: Digest
username="alice_private@atlanta.com",
realm="atlanta.com", nonce="",
uri="sip:registrar.atlanta.com", response=""
Security-Client: ipsec-3gpp; alg=hmac-sha-1-96; spi-
c=23456789; spi-s=12345678; port-c=2468; port-s=1357
Require: sec-agree
Proxy-Require: sec-agree
CSeq: 1 REGISTER
Content-Length: 0
```

The P-CSCF transfers the message REGISTER to the I-CSCF, including its IP address in the header `Path`. The P-CSCF can find the IP address of the I-CSCF by resolving the DNS (Domain Name System) on the basis of the domain name of Alice's UA.

The P-CSCF inserts the header `Require` containing the value `path` to ensure that the S-CSCF will take account of the header `Path`. If the S-CSCF does not support this header, it sends back a response `420 Bad extension`.

The P-CSCF also includes the header `P-Charging-Vector` whose parameter `icid` contains the charging ID.

The P-CSCF declares in the header `integrity-protected` that security has not been established with the UA.

The P-CSCF removes the headers `Require` and `Proxy-Require` which contained the `sec-agree` because IPSec security will be implemented by the P-CSCF.

```
REGISTER sip:registrar.atlanta.com SIP/2.0
Via: SIP/2.0/UDP
pcscf.atlanta.com;branch=z9hG4bK240f34.1
Via: SIP/2.0/UDP 192.0.2.101;branch=z9hG4bKnashds7
Max-Forwards: 69
P-Access-Network-Info: 3GPP-EUTRAN-FDD; eutran-cell-id-
3gpp=234151D0FCE11
Path: <sip:pcscf@192.0.2.111;lr>
Require: path
P-Charging-Vector: icid-
value="AyretyU0dm+6O2IrT5tAFrbHLso=023551024"
From: <sip:alice@atlanta.com>;tag=4fa3
To: <sip:alice@atlanta.com >
Contact: <sip:alice@192.0.2.101>;expires=600000
Call-ID: apb03a0s09dkjdfglkj49111
Authorization: Digest
username="alice_private@atlanta.com",
realm="atlanta.com", nonce="",
uri="sip:registrar.atlanta.com", response="", integrity-
protected="no"
CSeq: 1 REGISTER
Content-Length: 0
```

The I-CSCF contacts the HSS to retrieve the list of S-CSCFs which it is possible to assign to the UA. It selects an S-CSCF, to which it sends the REGISTER request.

The I-CSCF replaces the initial URI (`sip:registrar.atlanta.com`) with that of the S-CSCF (`sip:scscf.atlanta.com`).

```
REGISTER sip:scscf.atlanta.com SIP/2.0
Via: SIP/2.0/UDP
icscf.atlanta.com;branch=z9hG4bK351g45.1
Via: SIP/2.0/UDP
pcscf.atlanta.com;branch=z9hG4bK240f34.1
Via: SIP/2.0/UDP 192.0.2.101;branch=z9hG4bKnashds7
```

```
Max-Forwards: 68
P-Access-Network-Info: 3GPP-EUTRAN-FDD; eutran-cell-id-
3gpp=234151D0FCE11
Path: <sip:pcscf1@192.0.2.111;lr>
Require: path
P-Charging-Vector: icid-
value="AyretyU0dm+6O2IrT5tAFrbHLso=023551024"
From: <sip:alice@atlanta.com>;tag=4fa3
To: <sip:alice@atlanta.com >
Contact: <sip:alice@192.0.2.101>;expires=600000
Call-ID: apb03a0s09dkjdfglkj49111
Authorization: Digest
username="alice_private@atlanta.com",
realm="atlanta.com", nonce="",
uri="sip:registrar.atlanta.com", response="", integrity-
protected="no"
CSeq: 1 REGISTER
Content-Length: 0
```

The S-CSCF contacts the HSS to retrieve the authentication data for the mobile, and responds to it with a message `401 Unauthorized`. The response is transmitted to the I-CSCF whose IP address is contained in the first `Via` header.

At this stage, the IP address of the S-CSCF is registered in the HSS, and that of the P-CSCF in the S-CSCF.

The authentication data of the mobile comprise a quintuplet:

– a RAND challenge sent to the mobile in the message `401 Unauthorized`;

– the result expected from the challenge XRES;

– the network authentication token AUTN transmitted to the mobile in the message `401 Unauthorized`;

– the integrity key (IK) and encryption key (CK) for the establishment of the security association IPSec between the mobile and the P-CSCF. These keys are transmitted only to the P-CSCF, which removes them from the response `401 Unauthorized` before transferring it to the UA.

```
SIP/2.0 401 Unauthorized
Via: SIP/2.0/UDP
icscf.atlanta.com;branch=z9hG4bK351g45.1
Via: SIP/2.0/UDP
pcscf.atlanta.com;branch=z9hG4bK240f34.1
Via: SIP/2.0/UDP 192.0.2.101;branch=z9hG4bKnashds7
```

```
From: <sip:alice@atlanta.com>;tag=4fa3
To: <sip:alice@atlanta.com >; tag=5ef4
Call-ID: apb03a0s09dkjdfglkj49111
WWW-Authenticate: Digest realm="registrar.atlanta.com",
nonce=base64(RAND + AUTN + server specific data),
algorithm=AKAv1-MD5,
ik="00112233445566778899aabbccddeeff",
ck="ffeeddccbbaa11223344556677889900"
CSeq: 1 REGISTER
Content-Length: 0
```

The I-CSCF transfers the response to the P-CSCF after having removed the header `Via` containing its own IP address.

The P-CSCF transfers the response to Alice's UA after having removed the header `Via` containing its IP address and the IK and CK from the header `WWW-Authenticate`.

In the header `Security-Server`, the P-CSCF indicates the parameters of the security association IPSec established with Alice's UA.

```
SIP/2.0 401 Unauthorized
Via: SIP/2.0/UDP 192.0.2.101;branch=z9hG4bKnashds7
From: <sip:alice@atlanta.com>;tag=4fa3
To: <sip:alice@atlanta.com >;tag=5ef4
Call-ID: apb03a0s09dkjdfglkj49111
WWW-Authenticate: Digest realm="registrar.atlanta.com",
nonce=base64(RAND + AUTN + server specific data),
algorithm=AKAv1-MD5
Security-Server: ipsec-3gpp; q=0.1; alg=hmac-sha-1-96;
spi-c=98765432; spi-s=87654321; port-c=8642; port-s=7531
```

Upon receiving the message `401 Unauthorized`, the mobile authenticates the IMS network using the parameter AUTN and computes the response RES to the challenge RAND on the basis of a secret contained in the ISIM. It also computes the IK and CK.

5.2.2. *Second phase of registration*

The second phase of registration corresponds to the steps shown in Figure 5.8.

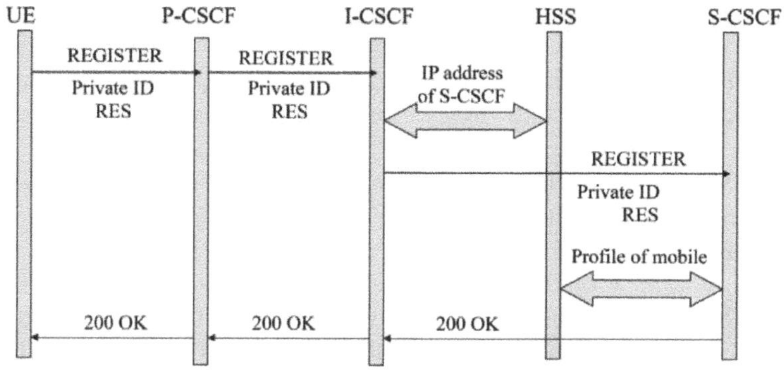

Figure 5.8. *Second phase of registration*

Alice's UA sends the P-CSCF a second REGISTER request containing its private ID and the response RES.

The REGISTER request, in the header `Security-Verify`, includes the security data received in the header `Security-Server` in the response `401 Unauthorized`.

```
REGISTER sip:registrar.atlanta.com SIP/2.0
Via: SIP/2.0/UDP 192.0.2.101;branch=z9hG4bKnashds7
Max-Forwards: 70
P-Access-Network-Info: 3GPP-EUTRAN-FDD; eutran-cell-id-
3gpp=234151D0FCE11
From: <sip:alice@atlanta.com>;tag=4fa3
To: <sip:alice@atlanta.com >
Contact: <sip:alice@192.0.2.101>;expires=600000
Call-ID: apb03a0s09dkjdfglkj49111
Authorization: Digest
username="alice_private@atlanta.com",
realm="registrar.atlanta.com", nonce=base64(RAND + AUTN
+ server specific data), algorithm=AKAv1-MD5,
uri="sip:registrar.atlanta.com",
response="6629fae49393a05397450978507c4ef1"
Security-Client: ipsec-3gpp; alg=hmac-sha-1-96; spi-
c=23456789; spi-s=12345678; port-c=2468; port-s=1357
Security-Verify: ipsec-3gpp; q=0.1; alg=hmac-sha-1-96;
spi-c=98765432; spi-s=87654321; port-c=8642; port-s=7531
CSeq: 2 REGISTER
Require: sec-agree
Proxy-Require: sec-agree
Content-Length: 0
```

The P-CSCF transfers the message REGISTER to the I-CSCF.

The I-CSCF contacts the HSS to retrieve the IP address of the S-CSCF and sends it the REGISTER request.

The S-CSCF compares the value of the RES received from the UA with that of the XRES received from the HSS. If the two values are identical, the mobile is authenticated.

The S-CSCF contacts the HSS to retrieve the profile of the mobile and responds to the UA with a message `200 OK` including its IP address in the header `Service Route`.

The registration is effective for a duration indicated in the parameter `expires` in the header `Contact`. The mobile has to renew its registration before expiration of that time, by way of the same procedure as for initial registration.

The S-CSCF, in the header `P-Associated-URI`, indicates the URIs registered implicitly, in addition to that indicated in the header `Contact`.

```
SIP/2.0 200 OK
Via: SIP/2.0/UDP
icscf.atlanta.com;branch=z9hG4bK351g45.1
Via: SIP/2.0/UDP
pcscf.atlanta.com;branch=z9hG4bK240f34.1
Via: SIP/2.0/UDP 192.0.2.101;branch=z9hG4bKnashds7
Path: <sip:pcscf1@192.0.2.111;lr>
Service-Route: <sip:scscf@192.0.2.113;lr>
From: <sip:alice@atlanta.com>;tag=4fa3
To: <sip:alice@atlanta.com >
Contact: <sip:alice@192.0.2.101>;expires=600000
Call-ID: apb03a0s09dkjdfglkj49111
CSeq: 2 REGISTER
P-Associated-URI: <sip:+1-212-555-1111@atlanta.com;
user=phone>
Content-Length: 0
```

The response follows the opposite path to that taken by the REGISTER request, thanks to the header `Via`.

5.2.3. *Subscription*

The different exchanges relating to the establishment of subscription to the registration event service are illustrated in Figure 5.9.

Alice's UA generates an initial SUBSCRIBE request.

The IP address of the P-CSCF of the domain (`atlanta.com`), contained in the header `Route`, is learned when the mobile is configured.

The IP address of the S-CSCF of the domain (`atlanta.com`), contained in the header `Route`, is learned during the registration of Alice's UA (information gleaned from the header `Service Route`).

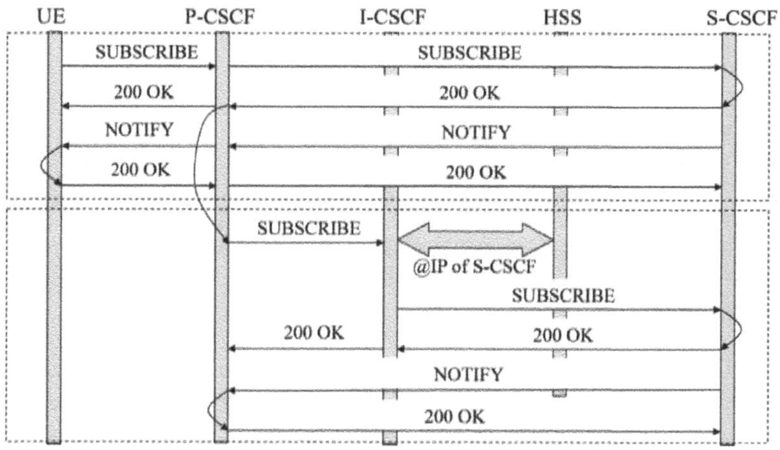

Figure 5.9. *Subscription to the event service*

If Alice's UA has multiple URIs, it indicates in the header `P-Preferred-Identity` that which is preferred (`sip:alice@atlanta.com`).

If Alice agrees for her identity to be presented, she indicates this in the header `Privacy` (value `none`).

Alice's UA defines in the header `Event` the type of event which it wishes to subscribe (value `reg` for registration).

Alice's UA publishes the duration of the subscription in the header `Expires`. The value must not be greater than that indicated during registration in the REGISTER request or in the response `200 OK`.

Alice's UA defines the format of the notifications of registration events in the header Accept (value application/reginfo+xml).

```
SUBSCRIBE sip:alice@atlanta.com SIP/2.0
Via: SIP/2.0/UDP 192.0.2.101:1357;branch=z9hG4bKnashds7
Max-Forwards: 70
Route: <sip:pcscf.atlanta.com:7531;lr>
Route: <sip:scscf.atlanta.com;lr>
P-Preferred-Identity: "Alice" <sip: alice@atlanta.com>
P-Access-Network-Info: 3GPP-EUTRAN-FDD; eutran-cell-id-3gpp=234151D0FCE11
Privacy: none
From: <sip:alice@atlanta.com>;tag=31415
To: <sip:alice@atlanta.com >
Call-ID: b89rjhnedlrfjflslj40a222
Require: sec-agree
Proxy-Require: sec-agree
CSeq: 61 SUBSCRIBE
Event: reg
Expires: 600000
Accept: application/reginfo+xml
Security-Verify: ipsec-3gpp; q=0.1; alg=hmac-sha-1-96;
spi-c=98765432; spi-s=87654321; port-c=8642; port-s=7531
Contact: <sip:292.0.2.101:1357>
Content-Length: 0
```

The P-CSCF removes the headers Route containing its own IP address and Security-Verify, and the value sec-agree. As the headers Proxy-Require and Require now contain no values, they are deleted.

The P-CSCF replaces the header P-Preferred-Identity with the header P-Asserted-Identity containing Alice's asserted URI.

The P-CSCF adds the headers Via, Record Route containing its own IP address and P-Charging-Vector. The header Record Route enables us to construct the route taken by subsequent requests.

The S-CSCF authorizes the subscription with the response 200 OK because it trusts the content of the header P-Asserted-Identity inserted by the P-CSCF and because Alice has subscribed to the registration event service. The response 200 OK is routed through the addresses contained in the headers Via.

Receipt of the message 200 OK triggers a subscription on the part of the P-CSCF in order to discover the state of registration of Alice's UA.

As the P-CSCF has not recorded any routing information during the registration phase, it does not know the IP address of the S-CSCF attributed to Alice's UA.

The request SUBSCRIBE from the P-CSCF is then transmitted to the I-CSCF, which consults the HSS to retrieve the IP address of the S-CSCF, and route the request accordingly.

5.2.4. *Notification*

5.2.4.1. *Register notification*

The different exchanges relating to the notification of registration events are shown in Figure 5.9.

The S-CSCF notifies Alice's UA of the registration of its URIs by sending a NOTIFY request.

The S-CSCF reports the state of the subscription in the header `Subscription-State` (value `active`) and its duration in the parameter `expires`.

```
NOTIFY sip:192.0.2.101:1357 SIP/2.0
Via: SIP/2.0/UDP
scscf.atlanta.com;branch=z9hG4bK332b23.1
Max-Forwards: 70
Route: <sip:pcscf.atlanta.com;lr>
From: <sip:alice@atlanta.com>;tag=31415
To: <sip: alice@atlanta.com>;tag=151170
Call-ID: b89rjhnedlrfjflslj40a222
CSeq: 42 NOTIFY
Subscription-State: active;expires=600000
Event: reg
Content-Type: application/reginfo+xml
Contact: <sip:scscf.atlanta.com>
Content-Length: (...)
```

The notification of registration elements is produced in an XML (eXtensible Markup Language) document of type `reginfo` attached to the SIP message.

```
<?xml version="1.0"?>
<reginfo xmlns="urn:ietf:params:xml:ns:reginfo"
                   version="1" state="full">
registration element
```

```
    <registration aor="sip: alice@atlanta.com" id="a7"
state="active">
        <contact id="76" state="active"
event="registered">
            <uri>sip:192.0.2.101</uri>
        </contact>
    </registration>
```

registration element

```
    <registration aor="tel: +1-212-555-1111" id="a8"
state="active">
        <contact id="77" state="active"
event="created">
            <uri>sip: 192.0.2.101</uri>
        </contact>
    </registration>
</reginfo>
```

Each registration element in the XML document includes the following attributes:

– the attribute `registration aor` (Address Of Record), followed by Alice's URI;

– the attribute `id`, followed by an identifier value;

– the attribute `state` of the registration. The attribute can be in the state `active` (the URI is registered) or `terminated` (the URI is de-registered).

Each contact of the registration element in turn contains the following attributes:

– the attribute `contact id`, followed by a contact identifier value;

– the attribute `state` of the contact of the element;

– the attribute which indicates the event which caused the change of state. This attribute can assume the following values:

- `registered`: registration has explicitly been carried out by a REGISTER request;

- `created`: registration has implicitly been carried out, following a REGISTER request;

- `deactivated`: registration has been canceled by the S-CSCF.

Similarly, the S-CSCF communicates the registration of Alice's UA to the P-CSCF by sending a NOTIFY request.

5.2.4.2. *De-register notification*

The different exchanges relating to the notification of de-registration events are shown in Figure 5.10.

De-registration can be initiated by the mobile or by the network. The mobile de-registers by sending a REGISTER request whose parameter `expires` in the header `Contact` has a value of zero.

This request triggers the transmission by the S-CSCF of a NOTIFY request to Alice's UA, on the one hand, and the P-CSCF on the other.

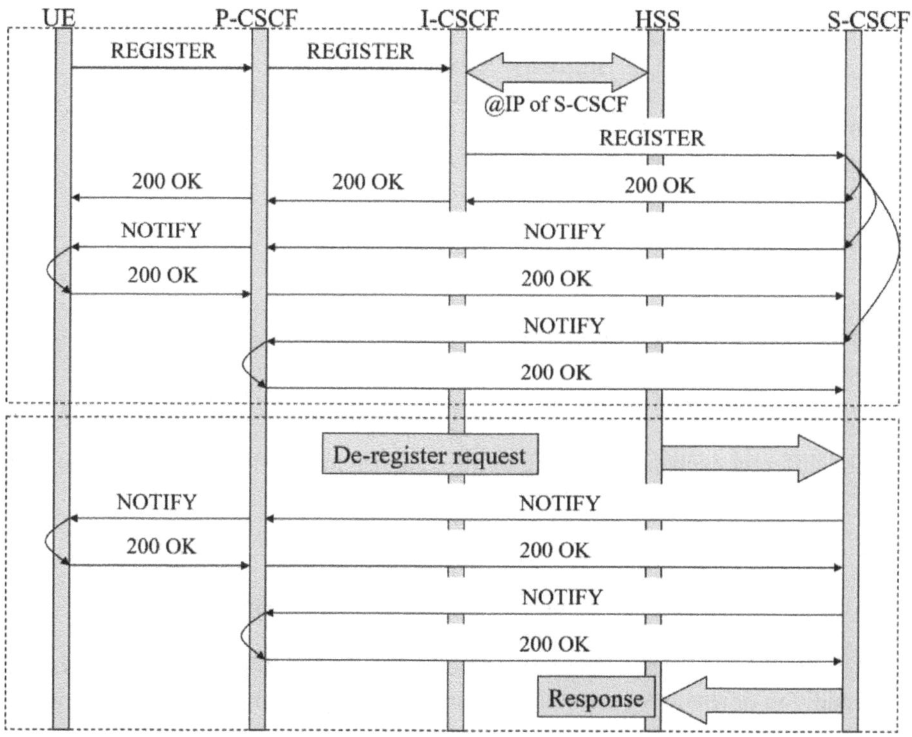

Figure 5.10. *De-register notification*

De-registration of Alice's UA by the network may take place on the initiative of the S-CSCF or of the HSS. As before, the S-CSCF sends the NOTIFY request respectively to Alice's UA and to the P-CSCF.

The de-registration of all of Alice's identities is specified in the body of the XML message associated with the NOTIFY request.

```
<?xml version="1.0"?>
<reginfo xmlns="urn:ietf:params:xml:ns:reginfo"
                 version="1" state="full">
     <registration aor="sip:alice@atlanta.com" id="as9"
                    state="terminated">
       <contact id="76" state="terminated"
event="deactivated">
           <uri>sip:192.0.2.101</uri>
       </contact>
     </registration>
     <registration aor=" tel: +1-212-555-1111" id="as11"
                    state="terminated">
       <contact id="77" state="terminated"
event="deactivated">
           <uri>sip: 192.0.2.101</uri>
       </contact>
     </registration>
</reginfo>
```

5.3. The session between IMSs

5.3.1. *Establishment of the session*

The different exchanges relating to the establishment of the session are shown in Figure 5.11.

5.3.1.1. *Initial INVITE request*

Alice's UA generates an initial INVITE request sent to Bob's URI.

Alice's UA specifies in the header `Require` that Bob's UA must support the pre-condition of resource reservation before activating the ringing of the telephone.

Alice's UA indicates in the header `Supported` that it supports the acknowledgement of provisional responses of `1xx` type (value `100rel`).

Alice's UA indicates in the header `Allow` the different supported methods (`INVITE, ACK, CANCEL, BYE, PRACK, UPDATE, REFER, MESSAGE`).

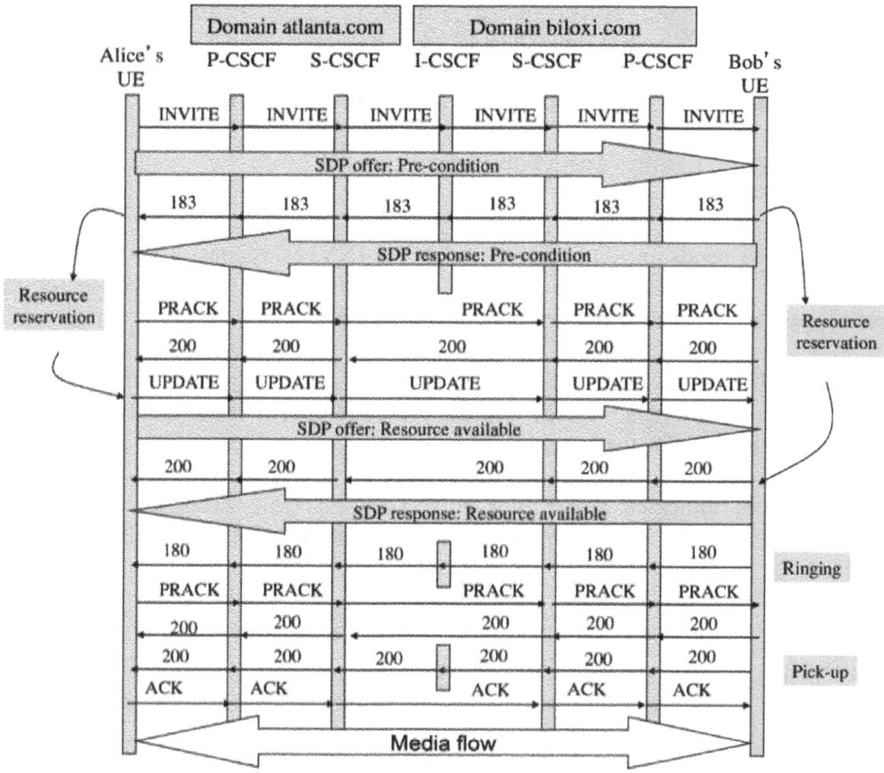

Figure 5.11. *Establishment of the session*

```
INVITE sip:bob@biloxi.com SIP/2.0
Via: SIP/2.0/UDP 192.0.2.101:1357;branch=z9hG4bKnashds7
Max-Forwards: 70
Route: <sip:pcscf.atlanta.com:7531;lr >
Route: <sip:scscf.atlanta.com;lr>
P-Preferred-Identity: "Alice" <sip:alice@atlanta.com>
P-Access-Network-Info: 3GPP-EUTRAN-FDD; eutran-cell-id-
3gpp=234151D0FCE11
Privacy: none
From: <sip:alice@atlanta.com>;tag=171828
To: <sip:bob@biloxi.com>
Call-ID: cb03a0s09a2sdfglkj490333
Cseq: 127 INVITE
Require: precondition, sec-agree
Proxy-Require: sec-agree
Supported: 100rel
```

```
Security-Verify: ipsec-3gpp; q=0.1; alg=hmac-sha-1-96;
spi-c=98765432; spi-s=87654321; port-c=8642; port-s=7531
Contact: <sip:192.0.2.101:1357>
Allow: INVITE, ACK, CANCEL, BYE, PRACK, UPDATE, REFER,
MESSAGE
Content-Type: application/sdp
Content-Length: (...)
```

Alice's UA sends an SDP offer in the initial INVITE request to Bob's UA. The offer lists all the types of media (audio, video) that Alice wishes to use for that session, and also lists the different codecs supported.

Alice's UA indicates in the SDP offer (`a=curr:qos local none, a=curr:qos remote none`) that the pre-condition of resource reservation has not been established for the local and remote UAs.

Alice's UA indicates in the SDP offer (`a=des:qos mandatory local sendrecv`) that the pre-condition of resource reservation is mandatory for the local UA.

Alice's UA indicates in the SDP offer (`a=des:qos none remote sendrecv`) that the pre-condition of resource reservation is unknown for the remote UA.

```
v=0
o=alice 2890844526 2890844526 IN IP4 192.0.2.101
s=-
c=IN IP4 192.0.2.101
t=0 0
m=video 3400 RTP/AVP 98 99
b=AS:75
a=curr:qos local none
a=curr:qos remote none
a=des:qos mandatory local sendrecv
a=des:qos none remote sendrecv
a=rtpmap:98 H263
a=fmtp:98 profile-level-id=0
a=rtpmap:99 MP4V-ES
m=audio 3456 RTP/AVP 97 96
b=AS:25.4
a=curr:qos local none
a=curr:qos remote none
a=des:qos mandatory local sendrecv
a=des:qos none remote sendrecv
a=rtpmap:97 AMR
a=fmtp:97 mode-set=0,2,5,7; mode-change-period=2
a=rtpmap:96 telephone-event
a=maxptime:20
```

The P-CSCF of the domain (atlanta.com) interacts with the PCRF to verify the parameters of the media types authorized by the EPS network. The PCRF examines the codecs contained in the SDP message carried by the INVITE request. If those codecs cannot be used by the EPS network, the P-CSCF sends Alice's UA a negative response 488 Not Acceptable Here. This rejection must contain enough information to enable Alice's UA to initialize a new attempt with the authorized media parameters.

The P-CSCF removes the headers Route containing its own IP address and Security-Verify, and the value sec-agree. As the header Proxy-Require no longer has a value, it is deleted.

The P-CSCF replaces the header P-Preferred-Identity with the header P-Asserted-Identity containing Alice's asserted URI.

The P-CSCF adds the headers Via and Record Route containing its own IP address and P-Charging-Vector. The header Record Route enables us to construct the route to be taken by subsequent requests.

The P-CSCF transfers the request to the S-CSCF of the domain (atlanta.com) whose IP address is contained in the header Route.

The HSS has, during registration, provided the S-CSCF with the user profile, containing the types of media authorized by the service offered. Similarly, the S-CSCF examines the information contained in the SDP message carried by the INVITE request. If the S-CSCF finds that these do not conform to the service profile, it sends Alice's UA a negative response 488 Not Acceptable Here.

The S-CSCF of the domain (atlanta.com) removes the header Route containing its own IP address. It retrieves the destination domain name (biloxi.com) from Bob's URI and contacts a DNS server to retrieve the IP address of the I-CSCF of the domain (biloxi.com) (Figure 5.11).

The S-CSCF completes the header P-Charging-Vector by adding the domain name (atlanta.com) that originated the session.

The S-CSCF adds the headers Via and Record Route containing its own IP address and transfers the request to the I-CSCF of the domain (biloxi.com).

The I-CSCF of the domain (biloxi.com) contacts the HSS to retrieve the IP address of the S-CSCF of the domain (biloxi.com) which has registered Bob's UA.

The I-CSCF adds the header `Via`, does not add a `Record Route` and transfers the request to the S-CSCF of the domain (`biloxi.com`). The subsequent requests therefore do not pass through the I-CSCF.

The S-CSCF of the domain (`biloxi.com`) verifies that the types of media proposed by Alice's UA correspond to the services offered to Bob.

The S-CSCF adds a header `Record Route` containing its own IP address.

The S-CSCF modifies the initial request, replacing Bob's URI with his IP address. The link between the URI and the IP address was created at registration. The S-CSCF adds Bob's URI to the header `P-Called-Party-ID` so that Bob knows which URI the INVITE request refers to.

The S-CSCF transfers the request to the P-CSCF of the domain (`biloxi.com`) (Figure 5.11). The IP address of the P-CSCF attributed to Bob's UA was learnt at registration (information contained in the header `Path`).

The P-CSCF of the domain (`biloxi.com`) verifies with the PCRF that the codecs proposed by Alice's UA are supported by Bob's EPS network.

The P-CSCF adds a header `Record Route` containing its own IP address and transfers the request to Bob's UA, whose IP address is included in the URI of the request.

Bob's UA stores the different headers `Record Route`, which it will later use to route subsequent requests.

5.3.1.2. *Response 183 Session Progress*

Bob's UA sends a response `183 Session Progress` to Alice's UA. The response is routed on the basis of the `Via` headers received.

The response `183 Session Progress` contains the `Record Route` headers received. This enables Alice's UA to retrieve the IP addresses of the CSCFs that need to process the subsequent requests.

The response `183 Session Progress` supplements the header `To`, adding the parameter `tag` to it.

Bob's UA indicates the value `100rel` in the header `Require` to indicate that the response requires acknowledgement from Alice's UA. In order to correlate the acknowledgement with the response, Bob's UA inserts the header `RSeq`.

```
SIP/2.0 183 Session Progress
Via: SIP/2.0/UDP
pcscf.biloxi.com:5088;branch=z9hG4bK361k21.1
Via SIP/2.0/UDP scscf.biloxi.com;branch=z9hG4bK764z87.1
Via SIP/2.0/UDP icscf.biloxi.com;branch=z9hG4bK871y12.1
Via SIP/2.0/UDP scscf.atlanta.com;branch=z9hG4bK332b23.1
Via SIP/2.0/UDP pcscf.atlanta.com;branch=z9hG4bK240f34.1
Via SIP/2.0/UDP 192.0.2.101:1357;branch=z9hG4bKnashds7
Record-Route: <sip: pcscf.biloxi.com:5088;lr;>
Record-Route: <sip:scscf. biloxi.com;lr>
Record-Route: <sip:scscf.atlanta.com;lr>
Record-Route: <sip:pcscf.atlanta.com;lr>
P-Access-Network-Info: 3GPP-EUTRAN-FDD; eutran-cell-id-
3gpp=234151D0FCE11
Privacy: none
From: <sip:alice@atlanta.com>;tag=171828
To: <bob@biloxi.com>;tag=314159
Call-ID: cb03a0s09a2sdfglkj490333
Cseq: 127 INVITE
Require: 100rel, precondition
Contact: <sip:bob@192.0.2.201:8805>
Allow: INVITE, ACK, CANCEL, BYE, PRACK, UPDATE, REFER,
MESSAGE
RSeq: 9021
Content-Type: application/sdp
Content-Length: (...)
```

In the response `183 Session Progress`, Bob's UA gives an SDP response in which it chooses a type of media from those proposed by Alice's UA.

Bob's UA indicates in the SDP message of the `183 Session Progress` that resource reservation is also necessary on its part before establishing a session.

```
v=0
o=bob 2890844527 2890844527 IN IP4 192.0.2.201
s=-
c=IN IP4 192.0.2.201
t=0 0
m=video 10001 RTP/AVP 98
b=AS:75
a=curr:qos local none
a=curr:qos remote none
a=des:qos mandatory local sendrecv
a=des:qos mandatory remote sendrecv
a=rtpmap:98 H263
m=audio 6544 RTP/AVP 97 96
```

```
b=AS:25.4
a=curr:qos local none
a=curr:qos remote none
a=des:qos mandatory local sendrecv
a=des:qos mandatory remote sendrecv
a=rtpmap:97 AMR
a=fmtp:97 mode-set=0,2,5,7; mode-change-period=2
a=rtpmap:96 telephone-event
a=maxptime:20
```

Alice's UA sends the subsequent PRACK request to acknowledge the provisional response `183 Session Progress`. It indicates in `Route` headers the IP addresses of the CSCF handling the request, i.e. the P/C-CSCF of the domains (`atlanta.com`) and (`biloxi.com`). This request is acknowledged with the response `200 OK`.

5.3.1.3. *UPDATE request*

When Alice's UA has confirmation of the resource reservation, it indicates this to Bob's UA in an SDP offer contained in an UPDATE request.

When Bob's UA has confirmation of the resource reservation, it notifies Alice's UA in an SDP offer contained in the response `200 OK` to the UPDATE request.

When the resource reservations are effective at both ends, Bob's telephone can ring and a response `180 Ringing` is transmitted to Alice's UA, generating a ringback tone for her.

The receipt of the response `180 Ringing` triggers the transmission of the acknowledgement request PRACK to guard against the loss of the response `180 Ringing`.

When Bob picks up, a response `200 OK` is sent to Alice's UE, which acknowledges with the subsequent ACK request. The session is then established.

5.3.2. *Termination of the session*

The different exchanges relating to the termination of the session are shown in Figure 5.12.

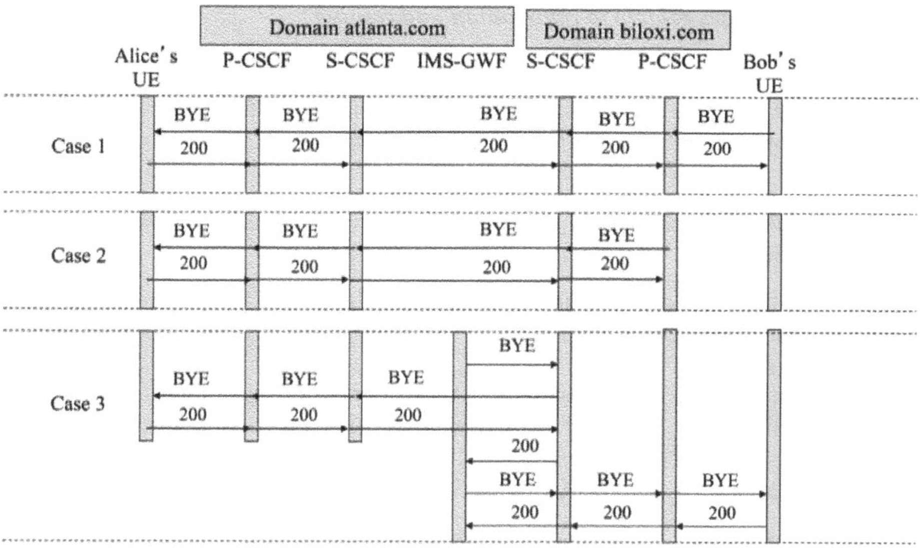

Figure 5.12. *Termination of the session*

The termination of the session can be triggered by the UA, P-CSCF or IMS-GWF, using the BYE method.

Termination of the session can be initiated by either UA when the communication has finished (Figure 5.12, case 1).

The function P-CSCF can also end the session if the mobile is no longer within radioelectric coverage (Figure 5.12, case 2). The PCRF sends this information to the P-CSCF by way of a DIAMETER message.

The IMS-GWF is an AS which controls the session in the case of pre-paid use. When the user's credit is exhausted, the IMS-GWF ends the session (Figure 5.12, case 3). This information can also be provided by the online charging system. Two BYE requests are needed to terminate the session: one request sent to Alice's UA and the second to Bob's UA.

5.4. DIAMETER messages

DIAMETER messages are exchanged between the CSCFs of the IMS network on the one hand and, on the other, either the HSS during registration of the UA or routing of the SIP request, or the PCRF for media control (Figure 5.2).

5.4.1. *The messages related to registration and routing*

The request UAR and the response UAA (User-Authorization-Request/Answer) are used between the I-CSCF and HSS during the two phases of registration of the UA. Upon receipt of a REGISTER request, the I-CSCF queries the HSS to retrieve:

– during the first phase, the list of S-CSCFs which it is possible to attribute to the UA (Figure 5.7);

– during the second phase, the IP address of the S-CSCF attributed to the UA (Figure 5.8).

The request MAR and response MAA (Multimedia-Auth-Request/Answer) are used between the S-CSCF and HSS during the first phase of registration. Upon receipt of the REGISTER request, the S-CSCF provides its IP address and retrieves the authentication data of the UA (Figure 5.7).

The request SAR and the response SAA (Server-Assignment-Request/Answer) are used between the S-CSCF and HSS during the second phase of registration. Upon receipt of the REGISTER request, the S-CSCF registers with the HSS and downloads the profile of the UA (Figure 5.8).

The request RTR and the response RTA (Registration-Termination-Request/Answer) are used between the HSS and S-CSCF when the HSS triggers the de-registration of the UA (Figure 5.10).

The request LIR and the response LIA (Location-Info-Request/Answer) are used between the I-CSCF and HSS for the routing of the INVITE request. With an incoming call, upon receipt of the INVITE request, the I-CSCF retrieves the identity of the S-CSCF attributed to the destination UA (Figure 5.11).

The request PPR and the response PPA (Push-Profile-Request/Answer) are used between the HSS and S-CSCF. These messages enable the HSS to notify a modification of the UA's profile.

5.4.2. *Messages relating to control of the media*

Messages relating to control of the media are exchanged between the P-CSCF and PCRF during the establishment of a session.

The request AAR (Authorization-Authentication-Request) is used by the P-CSCF to send the PCRF the characteristics of the media negotiated in the SDP message. The request also contains the parameter `icid` (charging identification) generated by the P-CSCF. The request is triggered by receipt of the response `183 Session Progress` (Figure 5.11).

The response AAA (Authorization-Authentication-Answer) from the PCRF acknowledges the request. If the codecs of the media are authorized, the following operations are triggered:

– the response AAA contains the parameter `gcid` for charging identification generated by the PDN GW of the EPS network. The parameters `icid` and `gcid` help to consolidate the bill, coupling charging units generated by the two networks IMS and EPS;

– the P-CSCF relays the response `183 Session Progress` (Figure 5.11);

– the PCRF triggers resource reservation at the level of the EPS network.

The STR (Session-Termination-Request) is used by the P-CSCF to inform the PCRF of the termination of the SIP session. This request is transmitted upon receipt of the BYE request (Figure 5.12). The STA (Session-Termination-Answer) from the PCRF acknowledges the request. On receiving this request, the PCRF releases the resources immobilized in the EPS network.

The ASR (Abort-Session-Request) is used by the PCRF to inform the P-CSCF that the resources reserved by the EPS network are no longer available. The ASA (Abort-Session-Answer) acknowledges the request. On receiving this request, the P-CSCF terminates the SIP session by sending a BYE request to each UA participating in the session (Figure 5.12).

5.5. Interoperation with the CS network

The CS (Circuit Service) network corresponds to a telephone connected to the PSTN or PLMN networks operating in CS mode. The interconnection between the IMS network and the CS network involves the BGCF (only for outgoing calls), MGCF, SGW and MGW.

5.5.1. *Call initiated by the IMS network*

The different exchanges relating to a call initiated by the IMS network are shown in Figure 5.13.

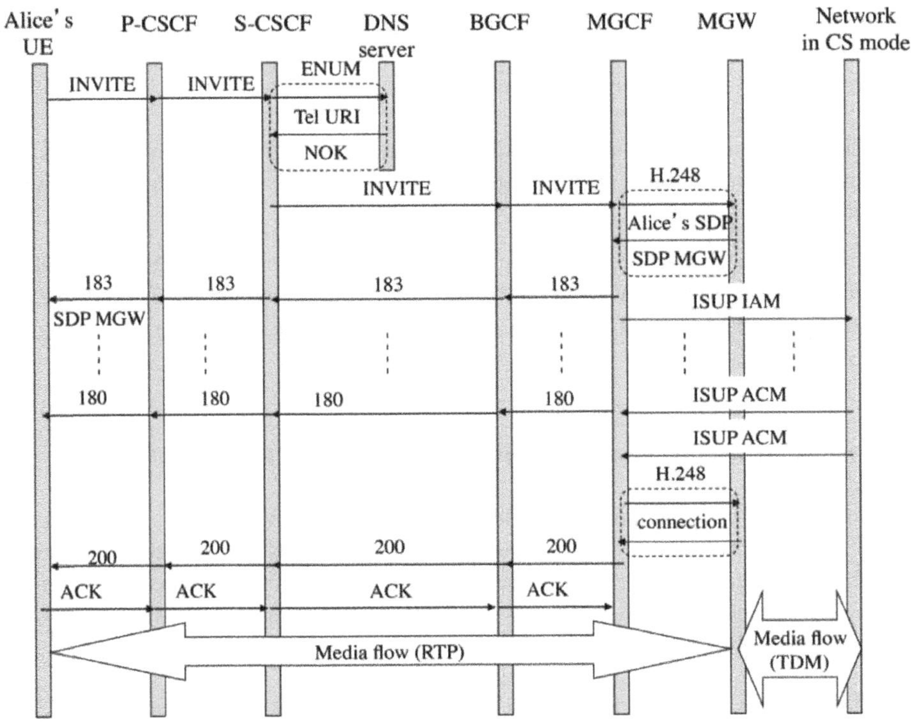

Figure 5.13. *Call initiated by the IMS network*

The call is generated by Alice's UA, by way of an INVITE request whose identity TEL URI contains the telephone number of the callee.

The S-CSCF carries out an ENUM resolution with the DNS server, to convert the TEL URI into a SIP URI. As the addressee is not a client of the IMS network, the response from the DNS server is negative.

The S-CSCF then routes the INVITE request to the BGCF, which consults a table to find the MGCF which provides the interconnection corresponding to the telephone address of the callee.

When the MGCF receives the INVITE request, it performs the following operations (Figure 5.13):

– it transfers the MGW the SDP message associated with the INVITE request;

– it commands the constitution of end points at the level of the MGW;

– it retrieves, in an SDP message, the characteristics of the media flow of the interface on the IP side of the MGW;

– it converts the INVITE request into an ISUP IAM message. This message is transmitted to the SGW using SIGTRAN, and then to the network in CS mode;

– it responds to Alice's UA with a message `183 Session Progress` in which the retrieved SDP message is inserted.

The network operating in CS mode reserves the resources and responds with an ISUP ACM message indicating that the call has been presented to the callee. The MGCF translates this ISUP message into a response `180 Ringing`. The ringback tone is applied to Alice's telephone.

When the callee hangs up, the MGCF receives the message ISUP ANM. It connects the end-points at the level of the MGW and sends the definitive response `200 OK` to Alice's UA (Figure 5.13). The ringback tone is then ceased and the communication is established.

5.5.2. *Call generated by the CS network*

The different exchanges relating to a call generated by the CS network are shown in Figure 5.14.

The call is generated by a CS entity, and results, at the level of the MGCF, in the receipt of the message ISUP IAM. The MGCF then performs the following operations:

– it creates the end-points with the MGW and retrieves the SDP message, describing the characteristics of the termination of the MGW on the IP side;

– it generates an INVITE request with the identity TEL URI containing the telephone number contained in the message ISUP IAM, associating the SDP message given by the MGW.

On receipt of the message `183 Session Progress`, the MGCF transfers the SDP message from the entity Bob to the MGW, thereby supplementing the characteristics of the end-point on the IP network.

When Bob's UA has confirmation of the resource reservation in the EPS network, the telephone rings and the MGCF receives the message `180 Ringing`. The MGCF generates the message ISUP ACM, sent to the CS network.

When Bob picks up, the MGCF receives the message `200 OK`. It connects the end-points of the MGW and transmits the message ISUP ANM to the network in CS mode.

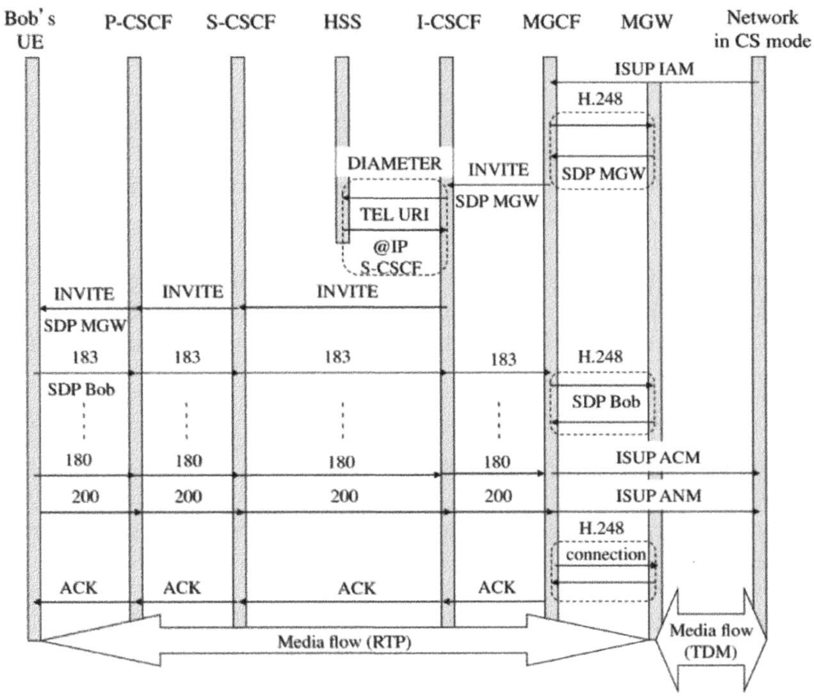

Figure 5.14. *Call initiated by the CS network*

5.5.3. *Release of the communication*

The different exchanges relating to the release of the communication are shown in Figure 5.15.

If the release of the communication is initiated by the CS, the MGCF receives the message ISUP REL from the CS network. The MGCF in turn generates a BYE request sent to Bob's UA and deletes the end-points of the MGW.

Upon receiving the BYE request, the P-CSCF triggers the release of the resources in the EPS network by way of the PCRF.

Bob's UA responds with the message `200 OK`. On receiving this message, the MGCF generates the message ISUP RLC, addressed to the CS network.

Bob's UA can also terminate the communication by generating the BYE request. Upon receiving the BYE request, the P-CSCF triggers the release of the resources of the EPS network through the PCRF, and the MGCF performs the following operations:

– it responds to Bob's UA with the message `200 OK`;

– it deletes the end-points of the MGW;

– it generates the message ISUP REL, sent to the CS network.

The CS network responds with the message ISUP RLC.

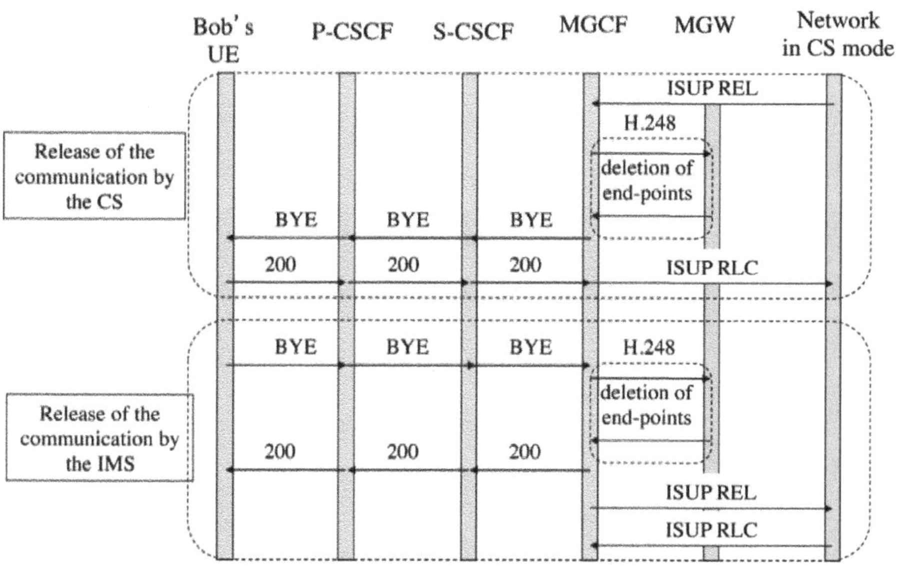

Figure 5.15. *Release of communication*

Chapter 6

Telephone Services

6.1. Service profile

When a user subscribes, they are assigned a private address, with which one or more service profiles are associated. Each service profile includes one or more public identities (SIP URI or TEL URI) and a service triggering logic in the form of iFC (initial Filter Criteria) (Figure 6.1).

The field *Core Network Service Authorization* in the iFC data structure contains an integer representing the type of media (audio, video) that the user has the right to negotiate in the SDP (Session Description Protocol) message (Figure 6.1).

The HSS (Home Subscriber Server) stores the data relating to the service profile of a user. The iFC data are transmitted to the S-CSCF (Serving – Call Session Control Function) when the user registers or when the service profile is modified.

The first field of iFC data is *Priority*, which determines the order in which the criteria need to be evaluated. The next field relates to the *Trigger Point*, which is a collection of filters (*Service Point Triggers*) (Figure 6.2).

Each filter consists of a logical operation carried out on the basis of the following parameters (Figure 6.2):

– the value of the URI (Uniform Resource Identifier) of the request (*Request-URI*);

– the type of method of the request (*SIP Method*);

– the content of the headers (*SIP Header*);

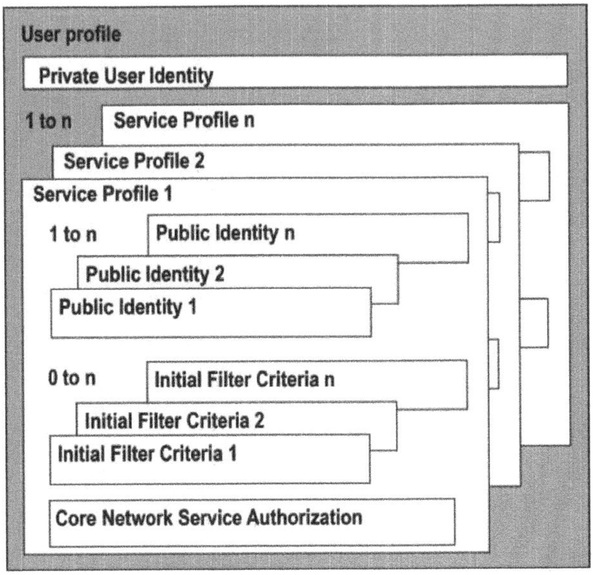

Figure 6.1. *Structure of service profile*

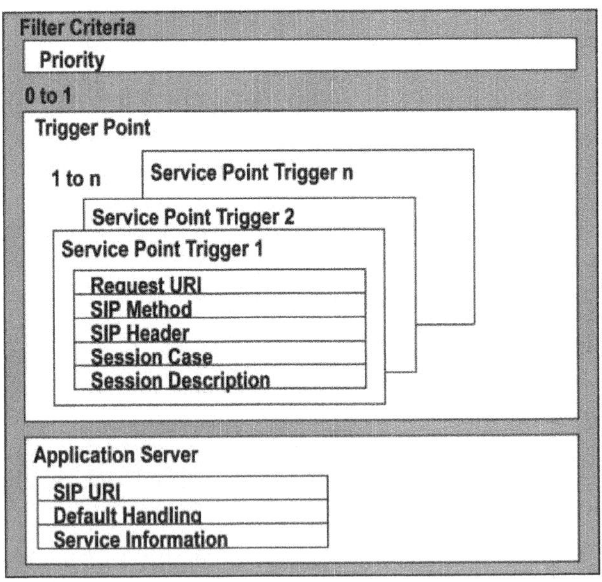

Figure 6.2. *Structure of iFC data*

– the type of call (outgoing or incoming), for a registered or unregistered mobile (*Session Case*);

– the content of the fields of the SDP message (*Session Description*).

The iFC data also contain the public identity of the TAS (Telephony Application Server) to which the S-CSCF transfers the request if the conditions are all fulfilled. The field *Default Handling* defines the treatment that needs to be applied to the request (continuation or cessation) if the S-CSCF cannot contact the TAS (Figure 6.2).

The field *Service Information* contains data which are transparent for the HSS and S-CSCF. These data are handled only by the TAS.

6.2. Communication Diversion

CDIV (Communication Diversion) consists of diverting a call (from Alice to Bob) to a different destination (Carol). In this scenario, CDIV is implemented by the TAS associated with the callee's (i.e. Bob's) S-CSCF.

6.2.1. *CFU*

CFU (Communication Forwarding Unconditional) consists of systematically forwarding an incoming call to another destination. However, the outgoing calls initiated by Bob's UA (User Agent) are not affected by this service.

Optionally, a subscription service may be offered to Bob, giving him an indication that CFU is active. This information is transmitted to him when Bob initializes a communication.

The different exchanges relating to CFU are shown in Figure 6.3.

The INVITE request transmitted by Alice's UA is received by Bob's S-CSCF. If the result of the iFC data handling is positive, the request is transmitted to the TAS which performs the following operations:

– it responds to Alice's UA with the `181 Call Is Being Forwarded`. This message alerts Alice's UA that the call has been transferred;

– it generates an INVITE request, sent to Carol's UA. This request includes the SDP message from the INVITE request received from Alice's UA.

When Carol's UA responds with the message `200 OK`, the TAS transfers the SDP message from Carol's UA in the message `200 OK` transmitted to Alice's UA.

The ACK request from Alice's UA (or respectively from the TAS) acknowledges receipt of the message `200 OK`. The media flow is then established between Alice's and Carol's UAs.

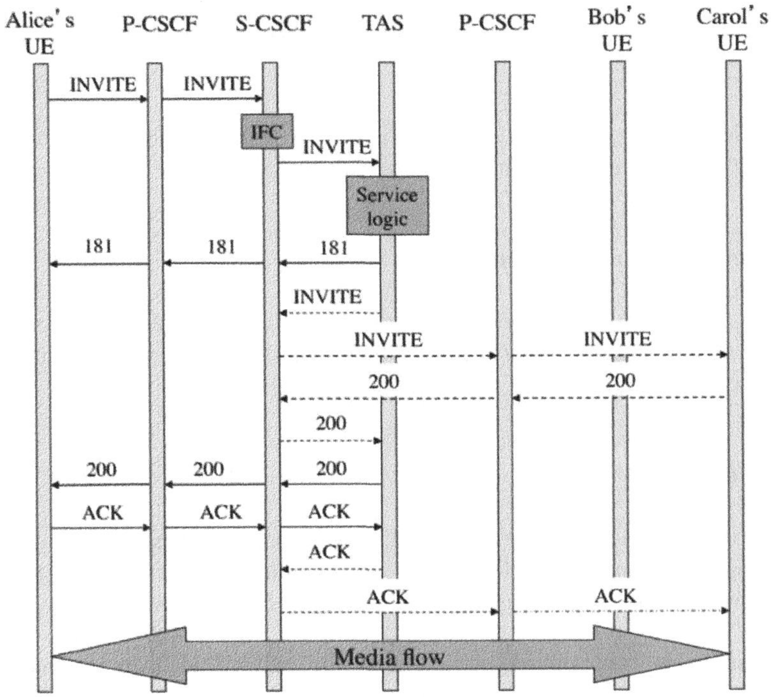

Figure 6.3. *CFU*

6.2.2. *CFB*

CFB (Communication Forwarding on Busy user) consists of forwarding an incoming call when the callee (Bob) is busy.

The different exchanges relating to CFB are illustrated in Figure 6.4.

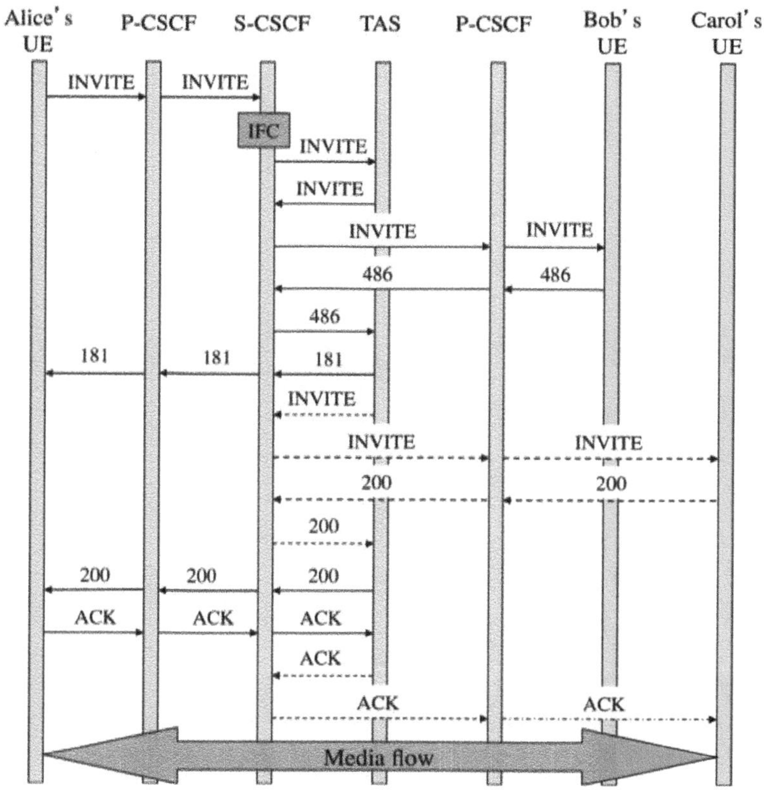

Figure 6.4. *CFB*

The INVITE request transmitted by Alice's UA to Bob's UA passes through the TAS associated with Bob's S-CSCF.

Bob's UA responds with the message `486 Busy Here`. This message is intercepted by the TAS, which implements communication forwarding to Carol's UA.

The following exchanges are identical to those shown for CFU.

6.2.3. *CFNR*

CFNR (Communication Forwarding on No Reply) involves forwarding an incoming call in the case of the lack of a response from the callee (Bob).

The different exchanges relating to CFNR are shown in Figure 6.5.

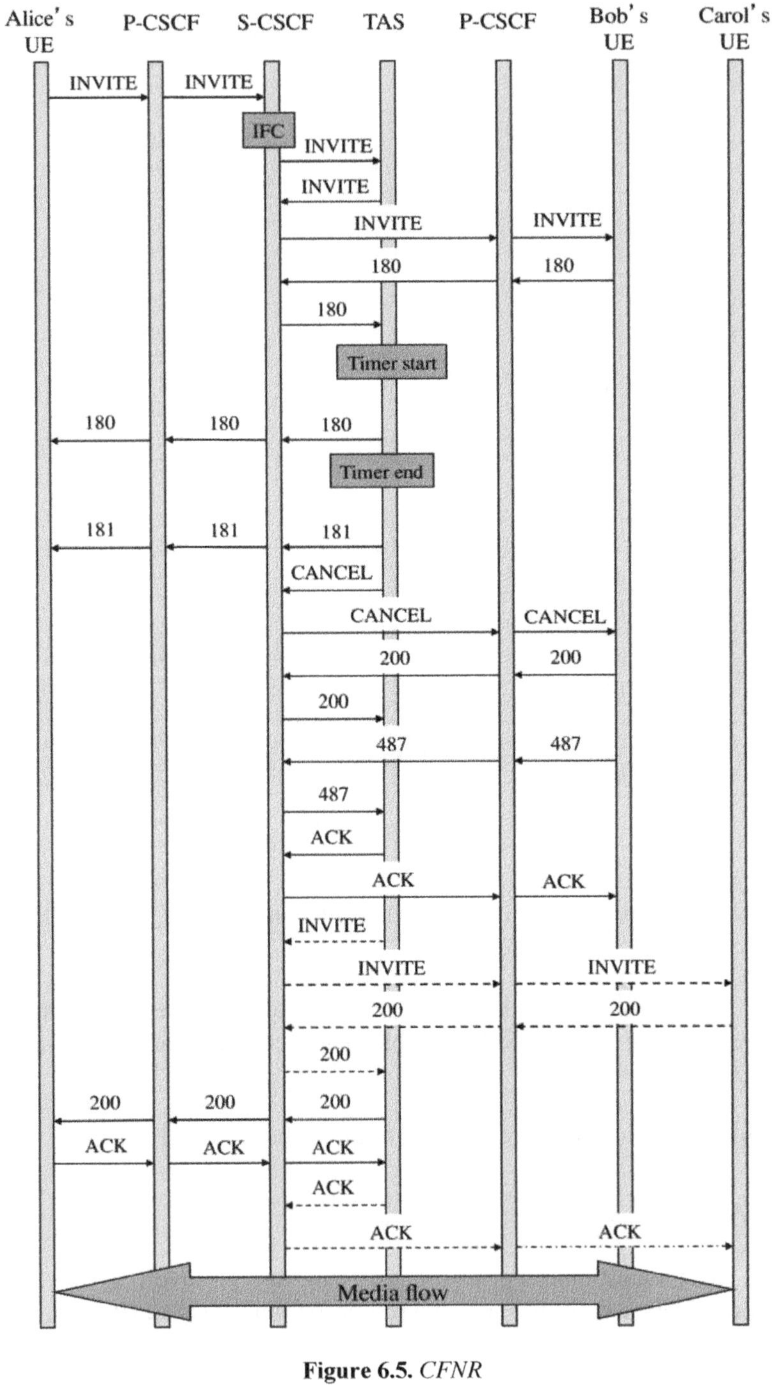

Figure 6.5. *CFNR*

On receiving the INVITE request, Bob's UA responds with the message `180 Ringing`. This message passes through the TAS, which starts a timer.

When that timer runs out, the TAS performs the following operations:

– it ends the dialog by sending the request CANCEL to Bob's UA, which triggers the sending of the response `487 Request Terminated`;

– it implements communication forwarding to Carol's UA.

6.2.4. *CD*

CD (Communication Deflection) enables the callee (Bob) to deflect a call if he does not want to answer the caller (Alice).

The different exchanges relating to CD are shown in Figure 6.6.

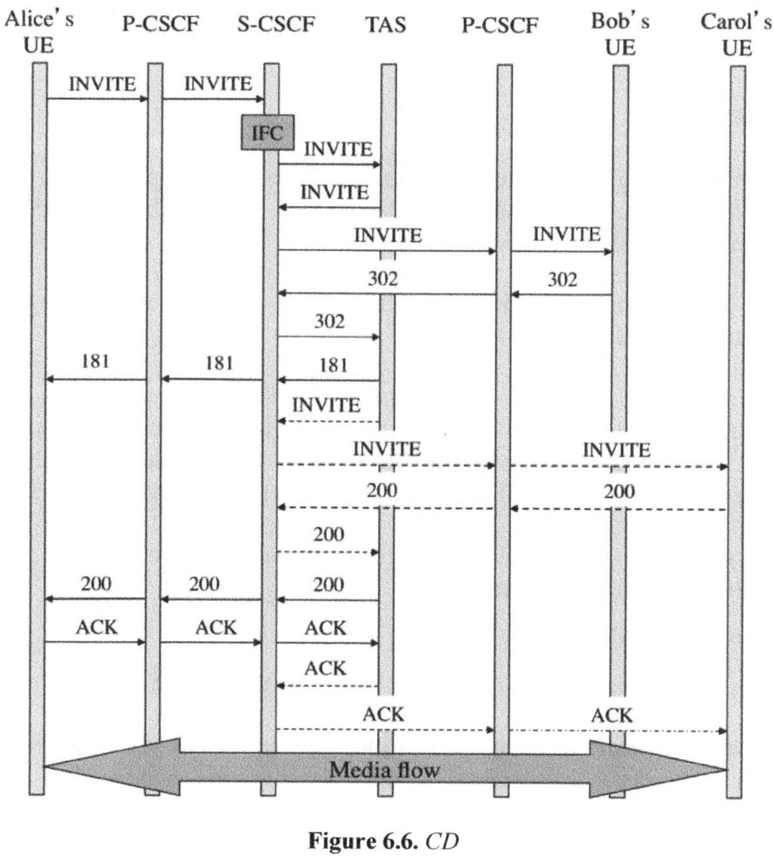

Figure 6.6. *CD*

The INVITE request sent by Alice's UA to Bob's UA passes through the TAS associated with Bob's S-CSCF.

Bob's UA responds with the message `302 Moved Temporarily`. The response contains Carol's URI, to which the call should be forwarded.

The response is intercepted by the TAS, which diverts the communication to Carol's UA.

6.2.5. *CFNL*

CFNL (Communication Forwarding on Not Logged-in) consists of diverting an incoming call if the callee (Bob) is not logged in with the S-CSCF.

The different exchanges relating to CFNL are identical to those for CFU.

6.3. Identification presentation

6.3.1. *OIP and OIR*

OIP (Originating Identification Presentation) consists of presenting the callee (Bob) with the identification of the caller (Alice), asserted by the IMS network (IP (Internet Protocol) Multimedia Subsystem).

OIR (Originating Identification Restriction) consists of masking Alice's identification from Bob.

Alice's UA inserts into the INVITE request the header `P-Preferred-Identity`, containing the public ID recorded implicitly or explicitly. The P-CSCF (Proxy-CSCF) replaces this header with the header `P-Asserted-Identity`. The value of this header is presented to Bob's UA.

Alice's UA can temporarily mask the presentation of her identity. It has to perform the following operations:

– it has to fill in the header `From` with the following value:

`<sip:anonymous@anonymous.invalid>`;

– it has to indicate the value `id` in the header `Privacy` so that the identity present in the header `P-Asserted-Identity` is not presented to Bob.

If Alice has subscribed to OIR, her TAS has to insert the value `id` into the header `Privacy`. It also has to modify the value of the header `From` to erase Alice's identity.

If Alice has subscribed to OIR, it can raise the restriction temporarily by inserting the value `none` in the header `Privacy`.

If Bob has not subscribed to OIP or if the header `Privacy` has the value `id`, his TAS has to remove the header `P-Asserted-Identity` and modify the value of the header `From` and remove Alice's identity.

6.3.2. *TIP and TIR*

TIP (Terminating Identification Presentation) consists of showing Alice the identity of the callee (Bob, or Carol in the case of communication forwarding), certified by the IMS network.

TIR (Terminating Identification Restriction) consists of masking Bob or Carol's identity from Alice.

If Alice has not subscribed to TIP or if the messages received contain the header `Privacy` with the value `id`, her TAS has to remove the header `P-Asserted-Identity` and modify the value of the header `From`, deleting Bob's or Carol's identity.

The callee's UA (i.e. Bob's or Carol's) can temporarily mask the presentation of his/her identity by adding the value `id` to the header `Privacy` in the response.

If Bob (or Carol) has subscribed to TIR, their TAS has to insert the header `Privacy` with the value `id`.

If Bob (or Carol) has subscribed to TIR, they can temporarily raise the restriction by indicating the value `none` in the header `Privacy`.

6.4. Message Waiting Indication

MWI (Message Waiting Indication) enables the TAS to indicate that a communication has occurred and that a voicemail (for instance) has been recorded. In this case, the TAS fulfills the function of a voicemail messaging service.

An MWI is transmitted if the user has subscribed to the MWI service.

The different exchanges relating to subscription to the MWI service are shown in Figure 6.7.

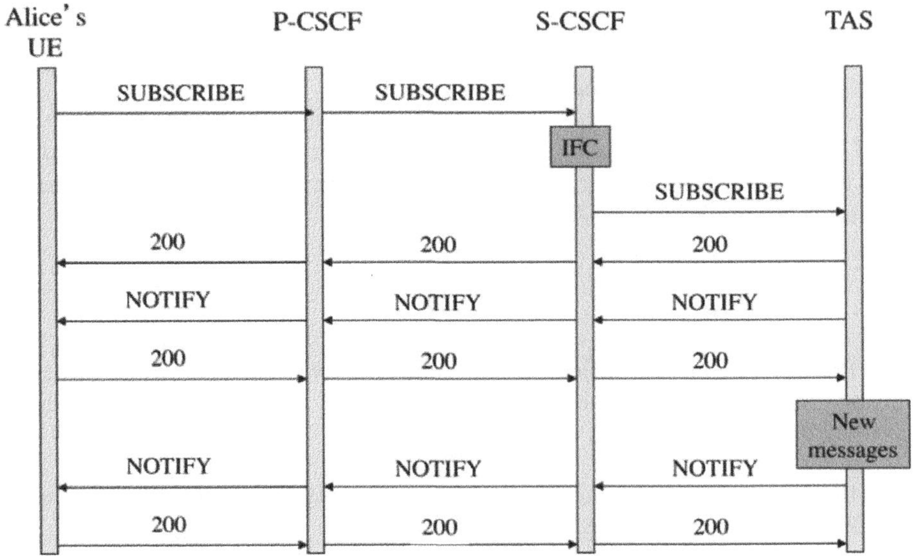

Figure 6.7. *Subscription to MWI service*

Alice's UA transmits a SUBSCRIBE request to subscribe to the MWI service. This message contains the header `event` with the value `message-summary`. It indicates in the header `Accept` the type of message body (value `application/simple-message-summary`).

The request is transmitted to the TAS by the S-CSCF following the iFC data processing.

The TAS verifies that the subscription for Alice's identity, contained in the header `P-Asserted-Identity`, is authorized and responds with a `200 OK`. Otherwise, the response is `403 Forbidden`.

When the subscription has been successful, the TAS immediately sends a NOTIFY request to synchronize the current state of the messages waiting. This initial notification contains only brief information about the recorded messages.

```
Messages-Waiting: yes
Message-Account: sip:alice@atlanta.com
Voice-Message: 2/1 (1/0)
```

```
Video-Message: 0/1 (0/0)
Fax-Message: 1/1 (0/1)
```

The message body of the NOTIFY request contains the following fields:

– `Message-Account`: this field includes Alice's public identity;

– `Voice-Message: 2/1 (1/0)`: this field shows the number of voice messages recorded:

- `2/1`: two new and one old voice messages are recorded;
- `(1/0)`: one of the two new messages is urgent.

– `Video-Message: 0/1 (0/0)`: this field shows the number of video messages recorded:

- `0/1`: one old message is recorded;
- `(0/0)`: no urgent messages are recorded.

– `Fax-Message: 1/1 (0/1)`: this field shows the number of fax messages recorded:

- `1/1`: one new and one old fax message are recorded;
- `(0/1)`: the old message is urgent.

If new messages are recorded, the TAS sends a new NOTIFY request whose message body provides a detailed description (the caller's identity, the subject of the message, the date) of the new messages.

In the example shown below, two urgent new voice messages and a new video message have been recorded.

```
Messages-Waiting: yes
Message-Account: sip:alice@atlanta.com
Voice-Message: 4/1 (3/0)
Video-Message: 1/1 (0/0)
Fax-Message: 1/1 (0/1)

To: <alice@atlanta.com>
From: <bob@biloxi.com>
Subject: call me back!
Date: 19 Apr 2005 21:45:31 -0700
Priority: urgent
Message-ID: 27775334485@mwi.atlanta.com
Message-Context: voice-message
```

```
To: <alice@atlanta.com>
From: <bob@biloxi.com>
Subject: Where are you that late???
Date: 19 Apr 2005 23:45:31 -0700
Priority: urgent
Message-ID: 27775334485@mwi.atlanta.com
Message-Context: voice-message

To: <alice@atlanta.com>
From: <carol@chicago.com>
Subject: Did you see that penalty!!!
Date: Tue, 19 Apr 2005 22:12:31 -0700
Priority: normal
Message-ID: 26775334485@mwi.atlanta.com
Message-Context: video-message
```

6.5. Call parking

The service of call parking enables a user to suspend an active communication and come back to it later on.

The different exchanges relating to call parking are shown in Figure 6.8.

Alice's UA sends a re-INVITE to Bob's UA to park the current call. Parking is achieved by modifying the attributes of the media flow in the SDP message:

– a=sendonly, if the media flow was previously in the state sendrecv;

– a=inactive, if the media flow was previously in the state recvonly.

Bob's UA responds positively with the message 200 OK. The SDP message is associated with the indication of the attribute a=recvonly.

Alice's UA sends a re-INVITE to Bob's UA to resume the parked communication. Recovery is achieved by modifying the attributes of the media flow in the SDP message:

– a=sendrecv, if the media flow was previously in the state sendonly;

– a=recvonly, if the media flow was previously in the state inactive.

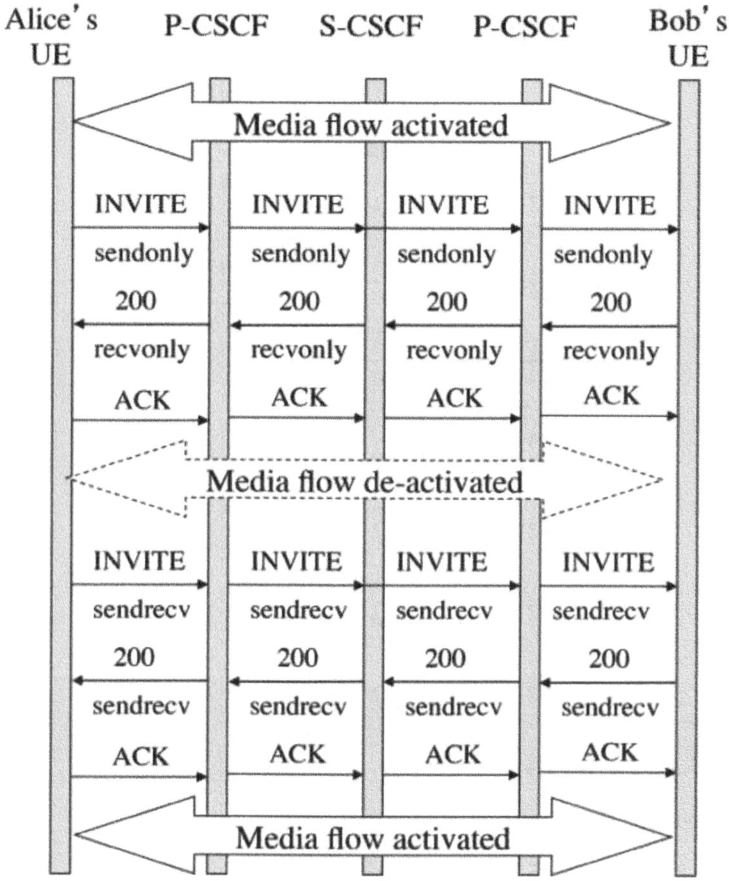

Figure 6.8. *Call parking service*

6.6. Conferencing

The conference service is provided by an AS comprising the entities MRFC (Multimedia Resource Function Controller) and MFRP (MRF Processor). The MFRP enables the media flows to be added together. The MRFC plays the part of a UA, and controls the MFRP.

The different exchanges relating to subscription to the conference service are shown in Figure 6.9.

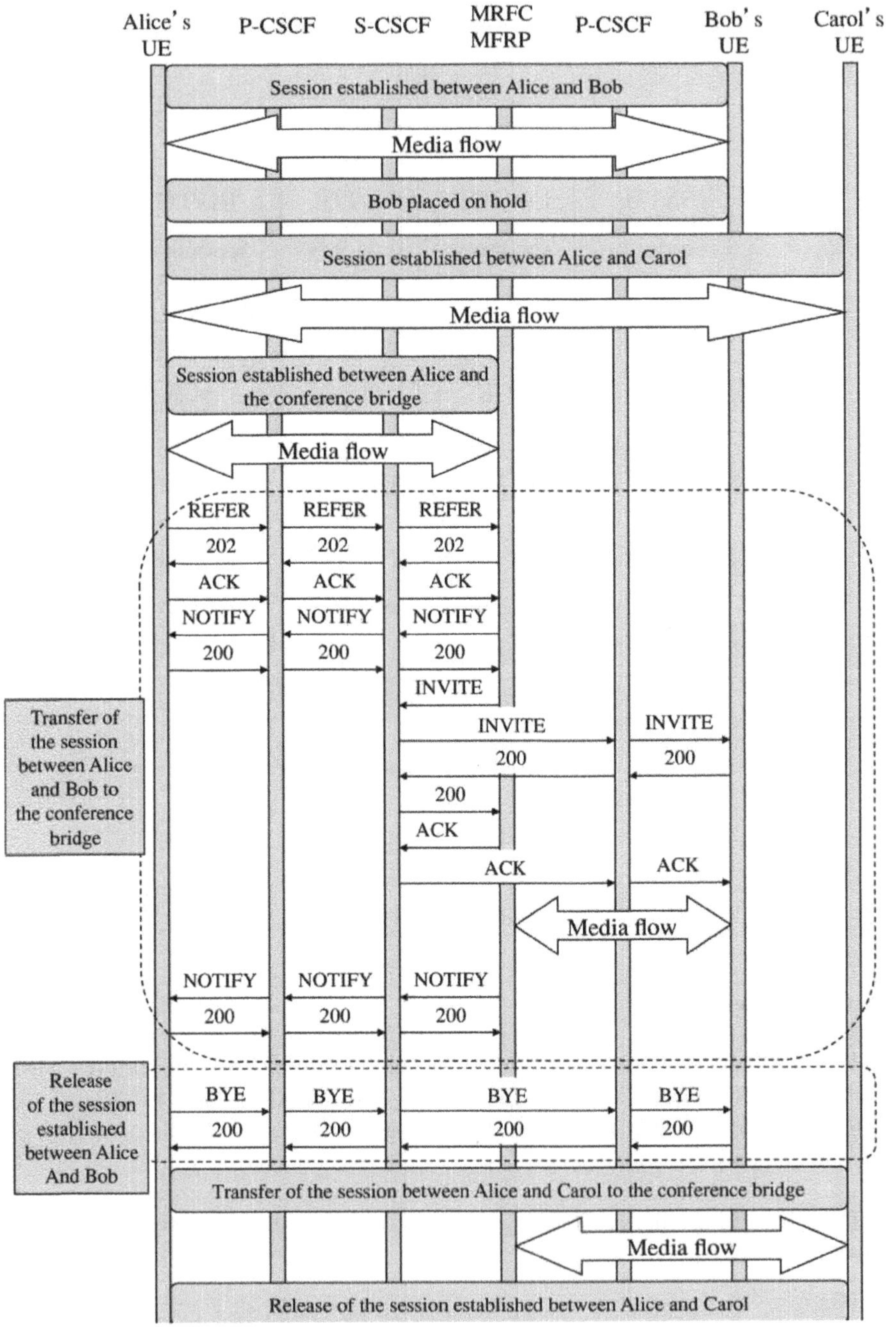

Figure 6.9. *Conferencing service*

Two UAs (Alice and Bob) are communicating. Alice decides to invite Carol and to activate the conference calling service. Alice puts Bob on hold, initializes a session with Carol, establishes a session with the conference bridge and transfers the two sessions established with Bob and Carol to the conference bridge.

When Alice's UA has established the sessions with Bob, Carol and the conference bridge, she begins the transfer of the session established with Bob to the conference bridge by sending the request REFER to the MRFC.

The REFER request includes the header `Refer-to` containing the references of the dialog initialized with Bob's UA (Bob's URI, header `Call-Id`, parameters `tag` in the headers `From` and `To`).

The MRFC sends a NOTIFY request to Alice's UA to indicate that the transfer has been activated. The header `Subscription-State` assumes the value `active`.

The MRFC initializes an INVITE request to Bob's UA, using the parameters communicated in the REFER request in the header `Replaces`. The session is established between Bob's UA and the conference bridge.

The MRFC sends a NOTIFY request to Alice's UA to communicate that the transfer has been terminated. The header `Subscription-State` assumes the value `terminated`.

Alice's UA sends a BYE request to Bob's UA to release the previously-established session.

6.7. Communication transfer

ECT (Explicit Communication Transfer) enables a participant (Bob) in a session (with Alice) to transfer the communication to a third party (Carol).

The three actors participating in the communication transfer service play the following roles:

– Alice's UA plays the role of *transferee*. It remains present in the communication that is the object of the transfer.

– Bob's UA plays the role of *transferor*. It initializes the transfer of the communication established with Alice.

– Carol's UA plays the role of *transfer target*. It replaces Bob's UA when the communication is transferred.

The communication will occur by one of two *modi operandi*:

– Bob transfers the communication to Carol, without having contacted her first (case 1).

– Bob transfers the communication to Carol, having first contacted her (case 2).

The different exchanges relating to communication transfer, for case 1, are shown in Figure 6.10.

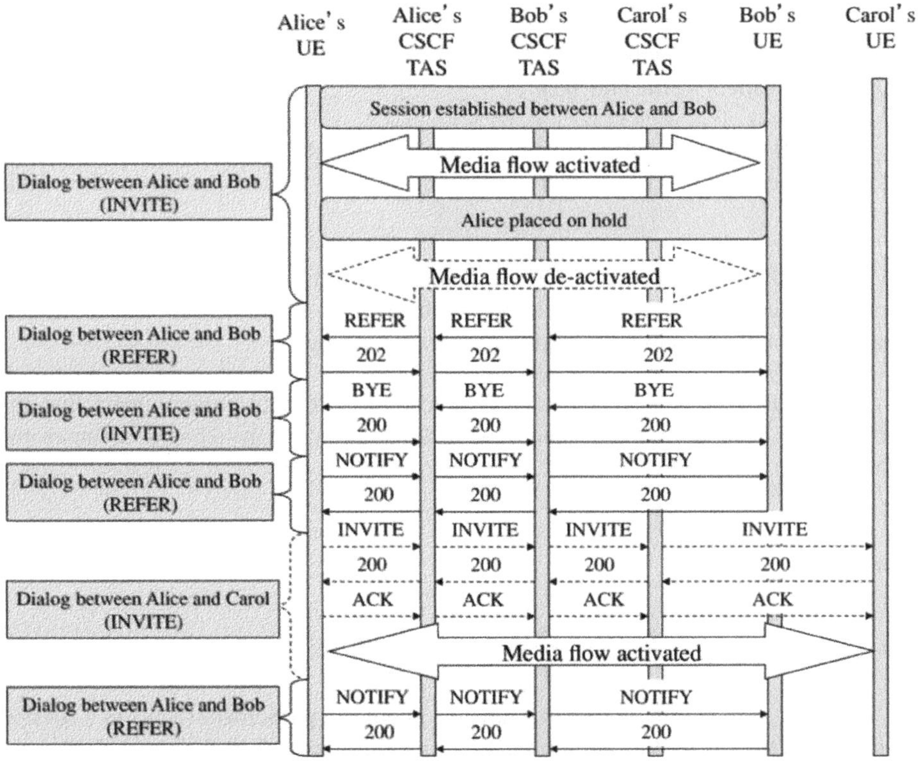

Figure 6.10. *ECT – case 1*

The communication transfer is initialized by Bob's UA by the sending of the REFER request. The header `Refer-To` contains Carol's URI.

Upon receiving this request, Bob's TAS checks that the destination (Carol's UA) is authorized. This arrangement is necessary because Bob's UA is charged for the communication between Bob and Carol (even if no session is really established).

Bob's TAS generates a specific session identity ECT URI which replaces Alice's URI in the header `Refer-To`. This arrangement enables Bob's TAS to retain control of the session.

As the REFER request is accepted, Bob's UA terminates the communication established with Alice's UA by the BYE request.

Alice's UA also informs Bob's UA, by way of a NOTIFY request, of the initialization of the session with Carol's UA.

Alice's UA initializes a new session whose destination corresponds to the ECT URI. Upon receipt of this request, Bob's TAS replaces this identity with that of Alice's UA.

When the session is established between Alice's and Carol's UAs, Alice's UA informs Bob's UA of this by way of a NOTIFY request.

The different exchanges relating to communication transfer, for case 2, are shown in Figure 6.11.

Two sessions are established: one between Alice and Bob, and the other between Bob and Carol. In the REFER request, Bob's UA indicates the characteristics of the dialog established with Carol.

Alice's UA uses these characteristics to initialize a dialog with Carol's UA. When the communication is transferred, the sessions between Alice and Bob on the one hand, and between Bob and Carol on the other, are released.

6.8. Communication Waiting

CW (Communication Waiting) gives information to the caller telling them that the callee is on a different communication but that he has been notified of the call.

Two possible scenarios may occur:

– If the TAS knows the callee's status and if the service is activated (case 1), it inserts the CW indication into the INVITE request sent to the callee.

– Otherwise (case 2), the TAS is informed of the situation in the response given by the callee.

The different exchanges relating to communication waiting, for case 1, are shown in Figure 6.12.

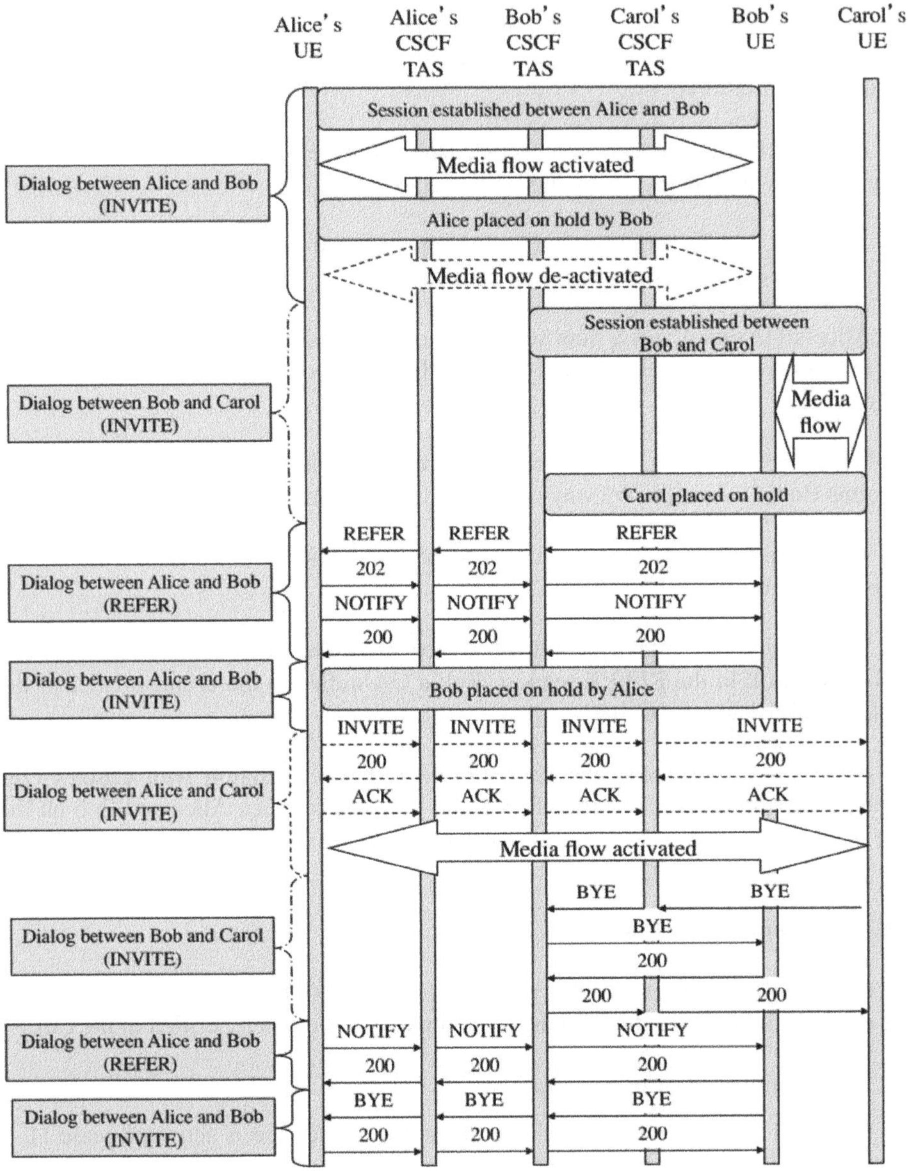

Figure 6.11. *Explicit communication transfer – case 2*

To the message body containing an SDP message (`application/sdp`) in the INVITE request, received from the S-CSCF, the TAS adds a second XML message (`application/vnd.3gpp.cw+xml`) giving the CW indication.

```
--boundary1
Content-Type: application/sdp

(SDP message)
--boundary1
Content-Type: application/vnd.3gpp.cw+xml

<?xml version="1.0"?>
<ims-cw xmlns="urn:3gpp:ns:cw:1.0">
      <communication-waiting-indication/>
</ims-cw>
--boundary1–
```

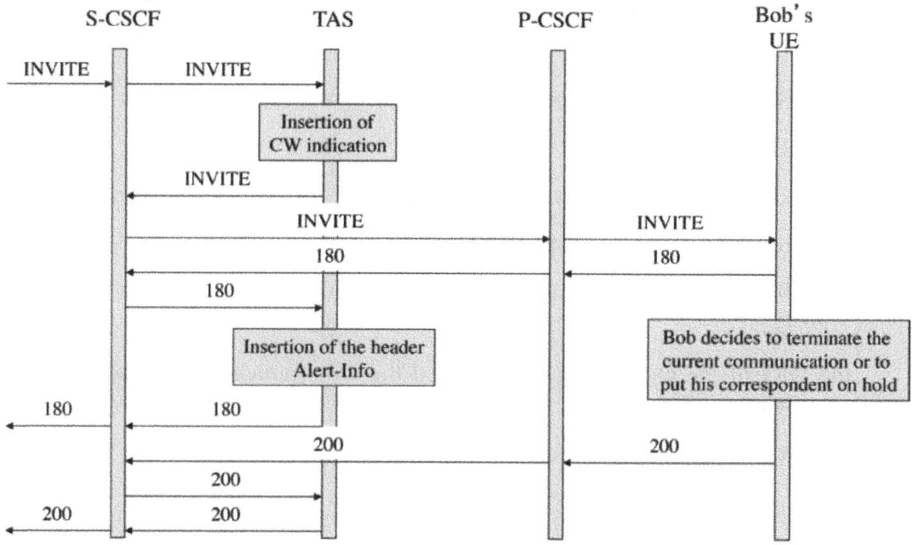

Figure 6.12. *CW service – case 1*

The TAS indicates in the header `Content-Type` that multiple messages appear in the message body (value `multipart/mixed`). It also includes in this header a parameter `boundary` corresponding to the boundary between the messages (value `boundary1`).

On receiving the message `180 Ringing`, the TAS implements the service CFNR, to transfer the call to a third party, in case Bob does not pick up the waiting communication.

To the message `180 Ringing` received from the P-CSCF, the TAS may optionally add the indication to the caller of a call waiting in the header `Alert-Info` (value `urn:alert:service:call-waiting`).

In case 2, the TAS simply relays the INVITE request, without enriching the message body. However, as before, it inserts the header `Alert-Info` into the response `180 Ringing`.

6.9. Malicious Communication Identification

MCID (Malicious Communication Identification) enables information relating to the identity of the source of the session to be tracked, upon request from the callee.

The TAS has to save the following elements from the INVITE request:

– the URI of the callee;

– the URI of the source, contained in the header `P-Asserted-Identity`;

– the date and time at which the INVITE request was received;

– the information relating to the communication transfer contained in the header `History-Info`;

– the URI of the entity which transferred the communication, contained in the header `Referred-By`;

– the URIs contained in the headers `Contact`, `To` and `From`.

If the INVITE request does not include these data, the TAS must send an INFO request to the network from which the request originates. The message body of this request contains the elements of identification of the session in XML format.

In turn, the TAS receives this information in an INFO request transmitted by the source network behind the request.

The UA of the callee can activate the service MCID. When the session has been established, it transmits a re-INVITE request with an XML message body `mcid`.

6.10. Automatic callback

6.10.1. *CCBS*

CCBS (Completion of Communications to Busy Subscriber) enables the caller (Alice) to avoid establishing a new session when the callee (Bob) is busy. Automatic callback occurs when the callee becomes free.

The different exchanges relating to CCBS are shown in Figure 6.13.

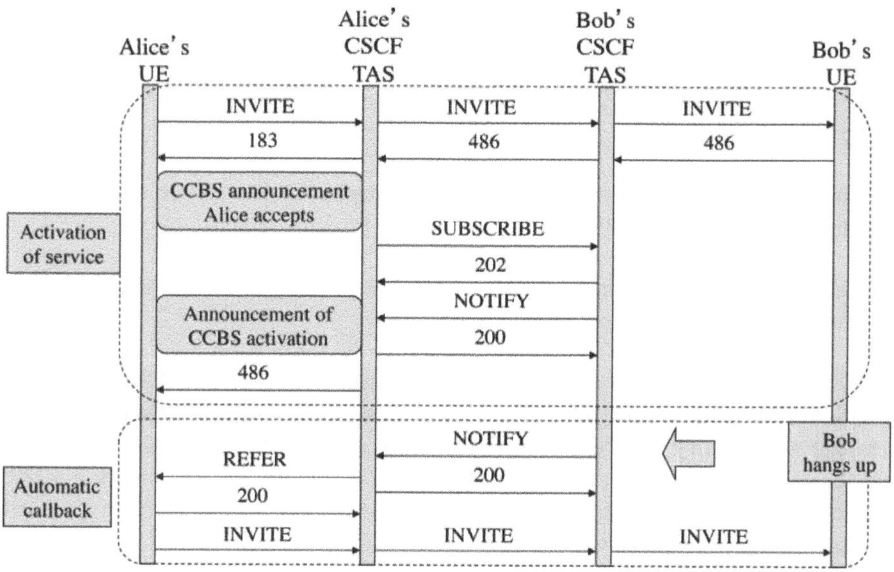

Figure 6.13. *CCBS*

Upon receiving the INVITE request, Bob's UA responds with the message `486 Busy Here`.

The response `486 Busy Here` is enriched with the header `Call-Info` by Bob's TAS. This header shows the following parameters:

– the URI of Bob's TAS;

– the service in question (parameter `purpose=call-completion`);

– the reason (m=BS Busy Subscriber).

```
SIP/2.0 486 Busy Here
Via:SIP/2.0/UDP tas.biloxi.com;branch= z9hG4bK332b23.1
```

```
Via:SIP/2.0/UDP scscf.biloxi.com;branch=z9hG4bK344a65.1
Via:SIP/2.0/UDP icscf.biloxi.com;branch=z9hG4bKj5hgrt2o
Via:SIP/2.0/UDP scscf.atlanta.com;branch=z9hG4bKehuehjgt
Via:SIP/2.0/UDP tas.atlanta.com;branch=z9hG4bKnashds7
Via:SIP/2.0/UDP pcscf.atlanta.com;branch=z9hG4bK240f34.1
Via: SIP/2.0/UDP
101.102.103.104:1357;branch=z9hG4bKnashds7
From: <sip:alice@atlanta.com>;tag=171828
To: <sip:bob@biloxi.com>;tag=314159
Call-ID: cb03a0s09a2sdfglkj490333
CSeq: 127 INVITE
Contact: <sip:201.202.203.204>
Content-Length: 0
Call-Info:<sip:tas.biloxi.com>;purpose=call-
completion;m=BS
```

The response `486 Busy Here` is momentarily held by Alice's TAS, for the time it takes to invoke CCBS.

The TAS sends the response `183 Session Progress` to Alice's UA and then an announcement to implement CCBS.

If the response from Alice's UA is positive, Alice's TAS generates a SUBSCRIBE request sent to Bob's TAS, whose URI had been recovered in the message `486 Busy Here`. The URI must be supplemented with the parameter `m` and its value `BS`). The `From` and `To` fields must contain Alice's and Bob's URIs respectively. The header `Event` specifies the type of event (`call-completion`) that is the subject of the subscription.

```
SUBSCRIBE sip:tas.biloxi.com;m=BS SIP/2.0
Via: SIP/2.0/UDP tas.atlanta.com;branch=z9hG4bKnashds7
Max-Forwards: 70
P-Asserted-Identity: <sip: alice@atlanta.com>
From: <sip:alice@atlanta.com>;tag=31415
To: <sip:bob@biloxi.com>
Call-ID: b89rjhnedlrfjflslj40a222
Call-Info:<sip:alice@atlanta.com>;purpose=call-
completion;m=BS
CSeq: 61 SUBSCRIBE
Event: call-completion
Expires: 2700
Contact: <sip:tas.atlanta.com>
Content-Length: 0
```

Bob's TAS sends Alice's TAS the notification of the subscription in the NOTIFY request. The message body states that Bob's UA is busy and the call-completion service is activated.

```
NOTIFY sip:tas.atlanta.com SIP/2.0
Via: SIP/2.0/UDP tas.biloxi.com;branch=z9hG4bK348923.1
Max-Forwards: 70
Route: <sip:scscf.biloxi.com>
P-Asserted-Identity: <sip:tas.biloxi.com>
From: <sip:bob@biloxi.com>;tag=151170
To: <sip:alice@atlanta.com>;tag=31415
Call-ID: b89rjhnedlrfjflslj40a222
CSeq: 42 NOTIFY
Subscription-State: active ;expires=2699
Event: call-completion
Contact: <sip:tas.biloxi.com>
Content-Type: application/call-completion
Content-Length: (...)

cc-state: queued
cc-service retention
```

When Bob hangs up, Bob's TAS informs Alice's TAS of this in the message body of the NOTIFY request.

```
cc-state: ready
cc-service retention
```

Alice's TAS initializes automatic call completing by sending to Alice's UA the REFER request. The parameter `m` (value `BS`) is associated with Alice's URI. Alice's UA once again calls Bob's UA by sending the INVITE request.

```
REFER sip:alice@atlanta.com;m=BS SIP/2.0
Via: SIP/2.0/UDP tas.atlanta.com;branch=z9hG4bK23273846
Max-Forwards: 70
Route: <sip:scscf.atlanta.com;lr>
P-Asserted-Identity: <sip:tas.atlanta.com>
From: <sip:tas.atlanta.com>; tag=161828
To: <sip:alice@atlanta.com>
Call-ID: cb03a0s09a2sdfglkj490333
Cseq: 127 REFER
Refer-To: <sip:bob@biloxi.com;method=INVITE>
Referred-By: <sip:tas.atlanta.com>
Contact: <sip:tas.atlanta.com>
Content-Length: 0
```

6.10.2. *CCNR*

The service CCNR (Completion of Communications on No Reply) enables the caller (Alice) to avoid initializing a new session when the callee (Bob) is not answering. Automatic callback occurs when Bob becomes free after having initialized any activity.

The different exchanges relating to CCNR are shown in Figure 6.14.

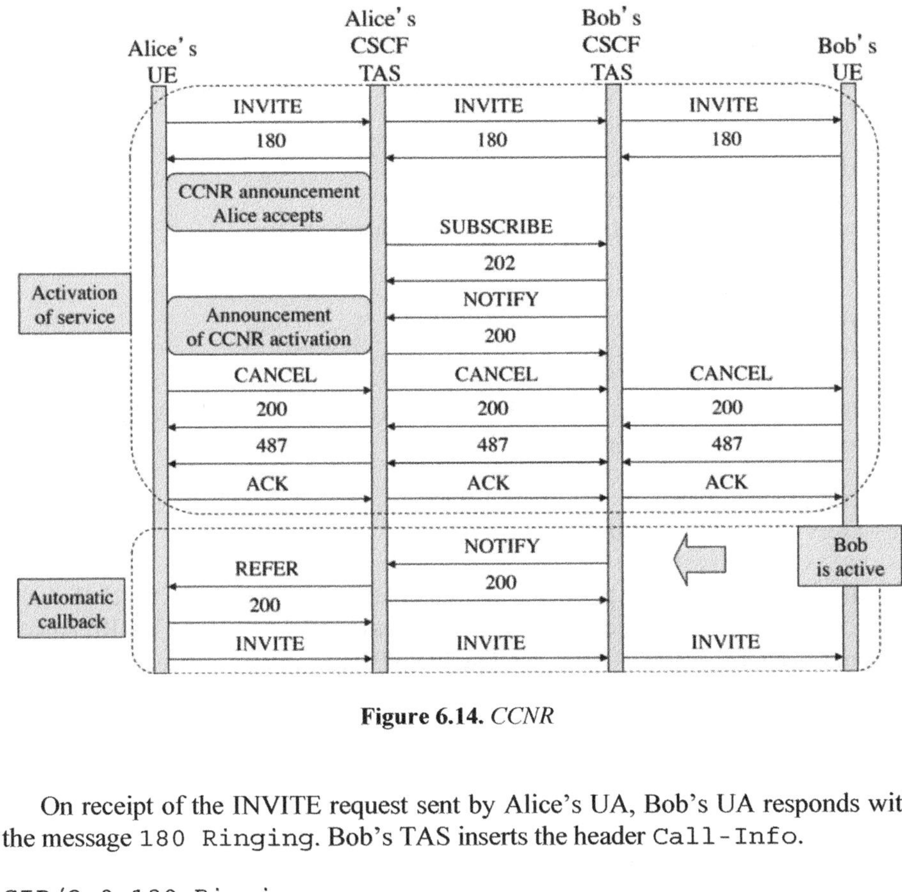

Figure 6.14. *CCNR*

On receipt of the INVITE request sent by Alice's UA, Bob's UA responds with the message `180 Ringing`. Bob's TAS inserts the header `Call-Info`.

```
SIP/2.0 180 Ringing
Via:SIP/2.0/UDP tas.biloxi.com;branch= z9hG4bK332b23.1
Via:SIP/2.0/UDP scscf.biloxi.com;branch=z9hG4bK344a65.1
Via:SIP/2.0/UDP icscf.biloxi.com;branch=z9hG4bKj5hgrt2o
Via:SIP/2.0/UDP scscf.atlanta.com;branch=z9hG4bKehuehjgt
Via:SIP/2.0/UDP tas.atlanta;branch=z9hG4bKnashds7
Via:SIP/2.0/UDP pcscf.atlanta.com;branch=z9hG4bK240f34.1
```

```
Via: 101.102.103.104:135;branch=z9hG4bKnashds7
From: <sip:alice@atlanta.com>;tag=171828
To: <sip:bob@biloxi.com>;tag=314159
Call-ID: cb03a0s09a2sdfglkj490333
CSeq: 127 INVITE
Retry-After: 3600
Contact: <sip:201.202.203.204>
Content-Length: 0
Call-Info:<sip:tas.biloxi.com>;purpose=call-
completion;m=NR
```

When CCNR is activated, Alice's UA terminates the session by sending the request CANCEL. The session is definitively terminated by the response `487 Request terminated`.

6.10.3. *CCNL*

The CCNL service (Completion of Communications on Not Logged-in) enables Alice to avoid initializing a new session when Bob is not logged in. Automatic callback occurs when the caller logs in.

The different exchanges relating to CCNR are shown in Figure 6.15.

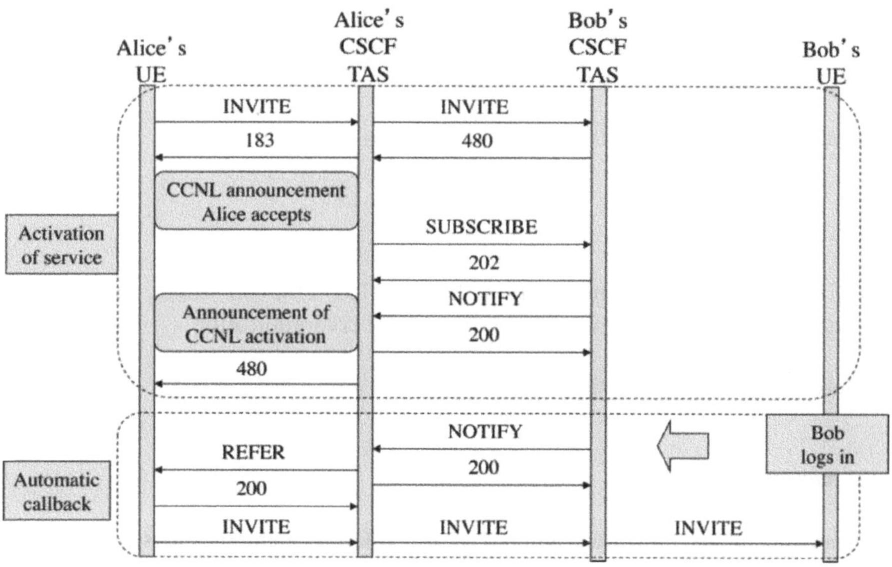

Figure 6.15. *CCNL*

Upon receipt of the INVITE request sent by Alice's UA, Bob's TAS responds with the message `480 Temporarily Unavailable` which inserts the header `Call-Info`.

Alice's TAS first holds the response and announces the CCNL service. The response is transmitted to Alice's UA when the service is activated.

6.11. Communication rejection

6.11.1. *ACR*

ACR (Anonymous Communication Rejection) enables Bob to reject an anonymous communication.

Bob's TAS rejects the communication because the INVITE request contains the header `Privacy` with the value `id`, `header` or `user`. It responds to the INVITE request with the message `433 Anonymity Disallowed`.

6.11.2. *ICB*

ICB (Incoming Communication Barring) enables the callee (Bob) to reject an incoming call belonging to a certain category of communication. The rule may include the URI of the caller, present in the header `P-Asserted-Identity` or `Referred-By` (in the case of call forwarding).

Bob's TAS responds with the `603 Decline`.

6.11.3. *OCB*

OCB (Outgoing Communication Barring) is able to block an outgoing communication belonging to a certain category of communication.

Bob's TAS responds with the message `603 Decline`.

6.12. Announcements

Announcements may be triggered at the establishment of the session, during communication or at the release of the session.

At the establishment of the session, an announcement can be generated by an AS in the following ways:

– using the header `Alert-Info` in the response `180 Ringing`. The header contains a URL (Uniform Resource Locator) which the receiver must contact to play the announcement;

– using the early media established between the UA and the AS, which uses the SDP message transmitted by the UA in the INVITE request. The AS comprises the two entities MRFC and MFRP.

The different exchanges relating to the announcement in the header `Alert-Info` are shown in Figure 6.16.

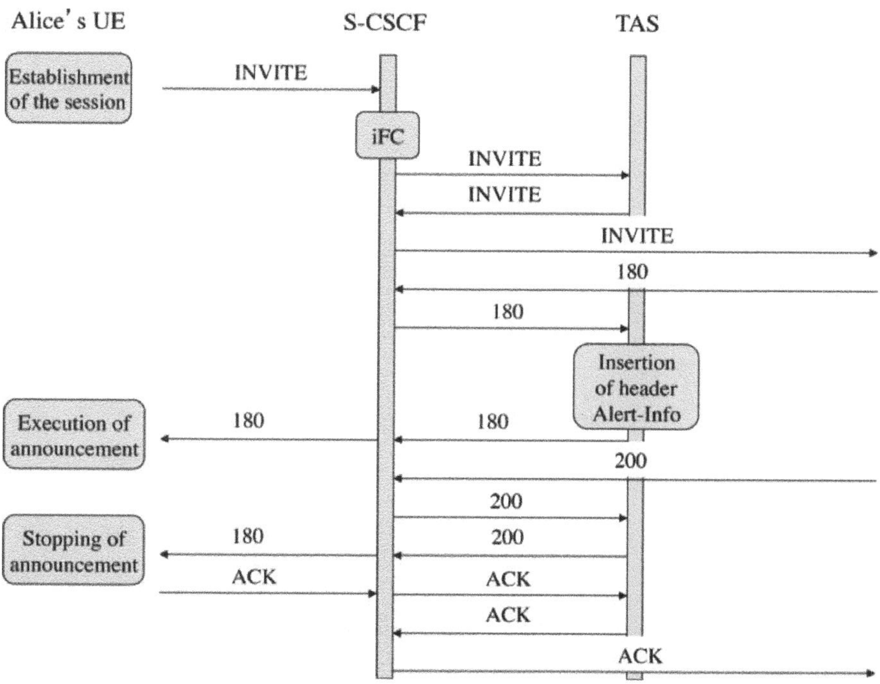

Figure 6.16. *The announcement in the header Alert-Info*

The S-CSCF receives the INVITE request from Alice's UA. The S-CSCF may belong to the domain of the caller (Alice) or the callee (Bob).

After having evaluated the conditions of the iFC, the S-CSCF transfers the request to the TAS.

The TAS sends the INVITE request back to the S-CSCF, which transfers it to the entities making up the callee's network.

The S-CSCF transfers the response `180 Ringing` to the TAS which adds the header `Alert-Info`. The response is then sent to Alice's UA.

Upon receiving the message `180 Ringing`, Alice's UA plays the announcement obtained from the URL in the header `Alert-Info`. This announcement is stopped on receiving the message `200 OK`.

The different exchanges relating to the announcement in the flow are shown in Figure 6.17.

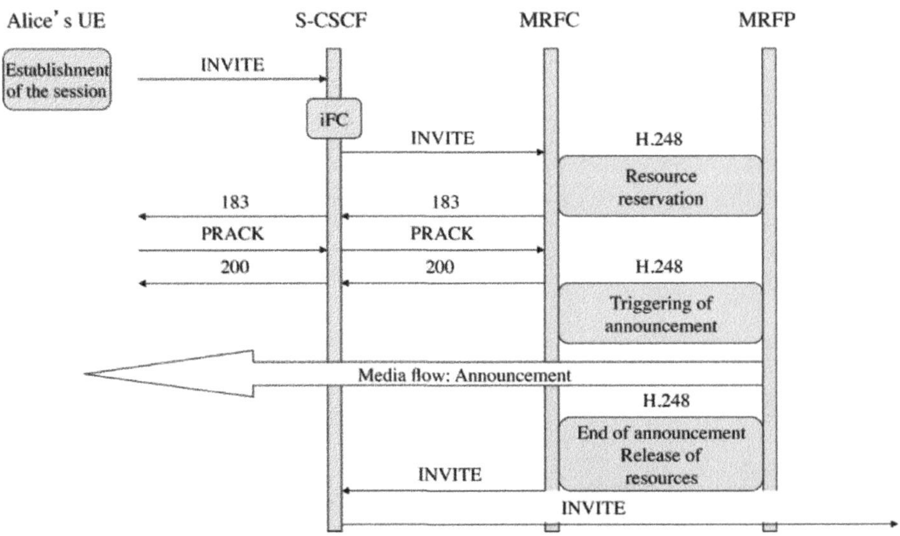

Figure 6.17. *The announcement in the media flow*

The MRFC momentarily holds the INVITE request received. It reserves the resources with the MRFP for the announcement and transfers the SDP message from Alice's UA to the MRFP.

The MRFC responds to Alice's UA with the message `183 Session progress`. This message includes the headers `P-Early-Media` (value `sendonly`) and `Require` (value `100rel`).

On receipt of the PRACK request, the MRFC triggers the announcement, in the MRFP, to be sent to Alice's UA.

When the announcement has been played, the MRFP informs the MRFC which, in return, releases the resources in the MRFP.

The MRFC sends the INVITE request to the S-CSCF which, in turn, transfers it to the entities in the callee's network.

During the establishment of the session, the announcement can also be generated by an AS following a rejected communication, in the following ways:

– using the header `Error-Info` in the response `3xx`, `4xx`, `5xx` or `6xx`. As before, the header contains a URL which the addressee needs to contact to play the announcement;

– using the early media, as before.

When the session is established, the announcement is generated by an AS, in the following ways:

– using the header `Call-Info` in a re-INVITE request. As before, the header contains a URL which the receiver needs to contact to play the announcement;

– using the early media, which need to be re-negotiated by a re-INVITE request in order to transmit the announcement.

When the session is released, the announcement can be sent to the UA of the user who has not yet hung up, using the early media as before.

Chapter 7

The SRVCC Function

7.1. Impact on architectures

When the mobile has established a telephone communication over the EPS (Evolved Packet System) network, it is necessary to be able to transfer that communication to the UMTS (Universal Mobile Telecommunications System) or GSM (Global System for Mobile) networks in the case of loss of coverage on the EPS network.

The function of SRVCC (Single Radio Voice Call Continuity) enables the telephone communication to be anchored in the IMS (IP Multimedia Sub-system) network, when the mobile switches from a network in PS (Packet Service) mode to one in CS (Circuit Service) mode.

7.1.1. *Impact on mobile networks*

The MME (Mobility Management Entity) in the EPS is affected by SRVCC. It performs the following functions:

– it separates the bearer for voice data from the other bearers not carrying voice data;

– via the Sv interface, it initializes the SRVCC procedure for handover of the voice data to the target cell in the GSM or UMTS network;

– it coordinates the handover from PS to CS mode for the voice flow, and possibly the handover from PS to PS mode for the other flows.

The MSC (Mobile Switching Center) Server in the GSM or UMTS network is also affected by SRVCC. It performs the following functions:

– it ensures the availability of the resources in the GSM or UMTS network before executing the handover;

– it coordinates the execution of the handover and transfer of telephone communication;

– it initiates the telephone communication transfer procedure.

Telephone communication transfer involves the transferal of telephone signaling (Figure 7.1) and voice data (Figure 7.2).

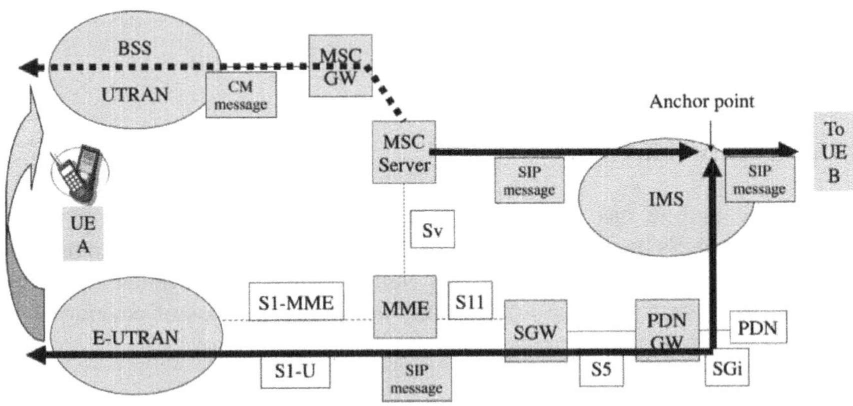

Figure 7.1. *Transfer of telephone signaling*

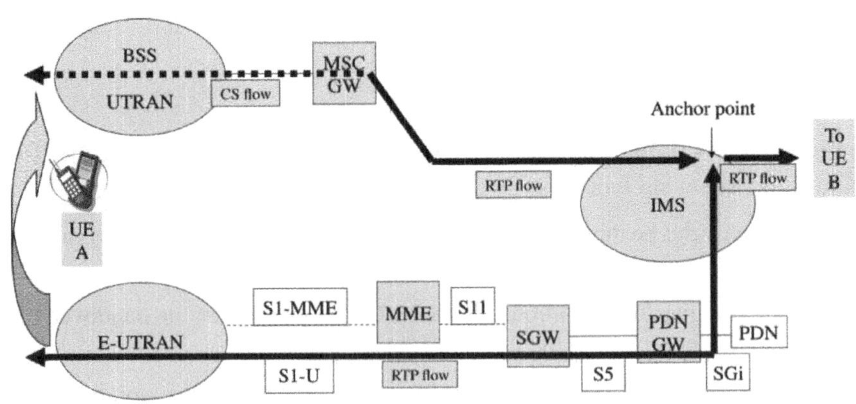

Figure 7.2. *Transfer of voice data*

The transfer of telephone signaling involves the transferal of the SIP (Session Initiation Protocol) signaling exchanged between the mobile and an anchor point in the IMS network, to the signaling made up of the CM (Call Management) protocol exchanged between the mobile and the MSC Server, and the SIP protocol exchanged between the MSC Server and the anchor point in the IMS network (Figure 7.1).

Transfer of voice data involves the transferal of the RTP (Real-Time Protocol) flow established between the mobile and an anchor point, to a junction of the CS bearer between the mobile and the MSC GW (MSC Gateway) and the RTP flow established between the MSC GW and the anchor point (Figure 7.2).

7.1.2. *Impact on the IMS network*

The SRVCC mechanism, at the level of the IMS network, results in the introduction of three entities (Figure 7.3):

– the SCC AS (Service Centralization and Continuity Application Server) which is in control of the SRVCC mechanism;

– the ATCF (Access Transfer Control Function) which provides the anchor point for the SIP signaling. The ATCF is introduced in the path of the signaling between the CSCF (Call Session Control Function) entities – on the one hand the P-CSCF (Proxy-CSCF) and on the other hand the I-CSCF (Interrogating-CSCF) or – S-CSCF (Serving-CSCF);

– the ATGW (Access Transfer Gateway) which provides the anchor point for the RTP flow. The ATGW is introduced in the path of the voice data.

The SRVCC mechanism brings modifications to the procedures of mobile registration and session establishment defined for the IMS network.

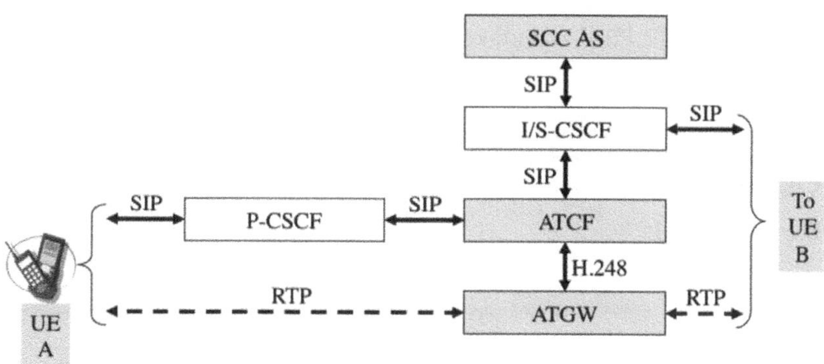

Figure 7.3. *Impact on IMS architecture*

7.1.2.1. *The SCC AS*

Following the registration of the mobile, the SCC AS receives a REGISTER message from the S-CSCF. This message contains the following information:

– the URI (Uniform Resource Identifier) of the ATCF;

– the PATH URI created by the ATCF. This code serves to identify the mobile during registration;

– the STN-SR (Session Transfer Number for SRVCC) created by the ATCF. This number serves to identify the session;

– the TEL URI of the mobile communicated by the S-CSCF (Serving Call Session Control Function).

It confirms registration of the mobile with the ATCF, and sends it the following data in a MESSAGE request:

– the PATH URI;

– its own identity SIP URI;

– the TEL URI of the mobile.

It updates the STN-SR in the HSS (Home Subscriber Server), which sends that STN-SR to the MME.

When the session is established, the SCC AS is in the path of the requests, following the result of the iFC (initial Filter Criteria) data processing, and of the responses. In the case of an outgoing call, the SCC AS is the first AS to be contacted. With an incoming call, the SCC AS is the last AS to be contacted.

When the communication is transferred, the SCC AS terminates the initial dialog in the P-CSCF (Proxy CSCF), ATCF and possibly in the UA (User Agent) if the latter has performed a PS-PS handover.

7.1.2.2. *The ATCF*

When the mobile registers, the ATCF receives the REGISTER request from the P-CSCF. It then generates the following information:

– the PATH URI;

– the session transfer number STN-SR.

Following registration, the ATCF receives a MESSAGE request from the SCC AS, confirming the registration of the mobile for SRVCC.

When the session is established, the ATCF anchors the RTP flow in the ATGW. In order to do so, it performs the following operations:

– it replaces Alice's SDP (Session Description Protocol) offer in the INVITE message with that of the ATGW;

– it replaces Bob's SDP offer in the response `183 Session Progress` with that of the ATGW.

When the communication is transferred, upon receiving the INVITE request from the MSC Server, the ATCF modifies the RTP flow, replacing the extremity Alice's UA with the extremity MSC GW. It also alerts the SCC AS to the transfer of the communication by initializing a new dialog.

7.1.2.3. *The ATGW*

The ATGW is controlled by the ATCF by way of the H.248 protocol. It is kept on the path of the RTP flow during the session. When the voice data are being transferred, it provides the anchor point for the RTP flow. It can also perform transcoding of the voice data if the codecs used in the EPS network are different from those used in the GSM or UMTS networks.

7.2. Procedures

7.2.1. *Registration*

The procedure of mobile registration is illustrated in Figure 7.4.

The registration procedure is initiated by the mobile. In the first phase of registration, the mobile sends the REGISTER message to the P-CSCF.

```
REGISTER sip:registrar.atlanta.com SIP/2.0
...
```

The P-CSCF transfers the REGISTER message to the ATCF, including its IP address in the header `Path`. In the headers `Route`, it indicates the addresses of the ATCF and I-CSCF (Interrogating CSCF).

```
REGISTER sip: registrar.atlanta.com SIP/2.0
...
```

```
Path: <sip:pcscf.atlanta.com;lr>
Route: <sip:atcf.atlanta.com;lr>
Route: <sip:icscf.atlanta.com;lr>
...
```

The ATCF generates a PATH URI. It transfers the REGISTER request to the I-CSCF, inserting the header `Path`, containing the PATH URI created, and the header `Feature-Caps` (Figure 7.4).

The header `Feature-Caps` contains the following parameters:

– `+g.3gpp.atcf`. This parameter shows the STN-SR;

– `+g.3gpp.atcf-mgmt`. This parameter contains the URI of the ATCF;

– `+g.3gpp.atcf-path`. This parameter contains the PATH URI generated by the ATCF;

– `+g.3gpp.srvcc`-alerting. This parameter indicates that the MSC Servers support SRVCC during the alerting phase.

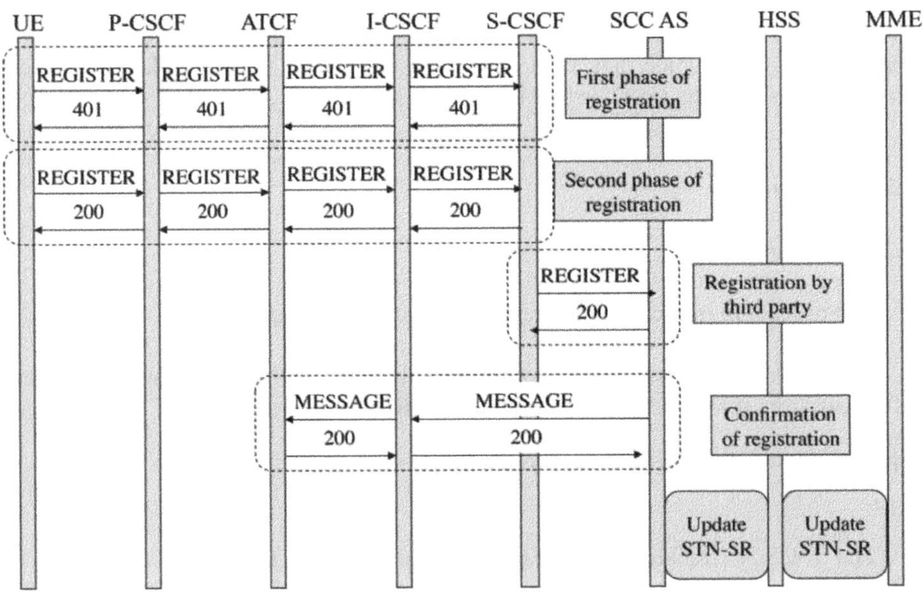

Figure 7.4. *Registration*

```
REGISTER sip:registrar.atlanta.com SIP/2.0
Feature-Caps:+g.3gpp.atcf="<tel:+1-237-888-9999>";
+g.3gpp.atcf-mgmt= "<sip:atcf.atlanta.com>";
+g.3gpp.atcf-path=
"<sip:termsdgfdfwe@atcf.atlanta.com>";
+g.3gpp.mid-call;+g.3gpp.srvcc-alerting
Path: <sip: termsdgfdfwe@atcf.atlanta.com;lr>
Path: <sip:pcscf.atlanta.com;lr>
Route: <sip:icscf.atlanta.com;lr>
...
```

In the second phase of registration, when the mobile has been authenticated, the S-CSCF responds with the message `200 OK`. This message copies the headers `Path` received in the REGISTER message.

```
SIP/2.0 200 OK
...
Path: <sip: termsdgfdfwe@atcf.atlanta.com;lr>
Path: <sip:pcscf.atlanta.com;lr>
...
```

The ATCF includes the header `Feature-Caps` with the parameter `+g.3gpp.atcf` containing the STN-SR.

```
SIP/2.0 200 OK
Feature-Caps:+g.3gpp.atcf="<tel:+1-237-888-9999>"
...
Path: <sip: termsdgfdfwe@atcf.atlanta.com;lr>
Path: <sip:pcscf.atlanta.com;lr>
...
```

The S-CSCF sends the SCC AS the state of registration of the mobile in a REGISTER message. The message body contains the first REGISTER request received by the S-CSCF and the response `200 OK`.

```
REGISTER sip: sccas.atlanta.com /2.0
Via: SIP/2.0/TCP scscf.atlanta.com;branch=z9hG499ffhy
Max-Forwards: 70
From: <sip:scscf.atlanta.com>; tag=538ya
To: <sip:alice@atlanta.com>
Call-ID: 1asdaddlrfjflslj40a222
```

```
Contact: <sip:scscf.atlanta.com>; expires=600000
CSeq: 87 REGISTER
Content-Type: multipart/mixed;boundary="boundary1"
Content-Length: (...)

--boundary1
Content-Type: message/sip

REGISTER sip:registrar.atlanta.com SIP/2.0
...

--boundary1
Content-Type: message/sip

SIP/2.0 200 OK
...

--boundary1—
```

The SCC AS sends the ATCF (`sip:atcf.atlanta.com`) the information relating to SRVCC in the body of a MESSAGE request. The message body is in XML (eXtensible Markup Language) format. It contains the following information:

– the PATH URI generated by the ATCF (`sip:termsdgfdfwe@atcf.atlanta.com`);

– the identity of the SCC AS (`sip:sccas.atlanta.com`);

– the TEL URI of the mobile (`tel:+1-237-555-1111`). This identity is present in the header `P-Associated-URI` in the response `200 OK`.

```
MESSAGE sip:atcf.atlanta.com SIP/2.0
Via: SIP/2.0/UDP
sccas.atlanta.com;branch=z9hG4bKnas588339
Max-Forwards: 70
From: <sip:sccas.atlanta.com>;tag=aassd
To: sip:atcf. atlanta.com
Call-ID: sdvasdfgfasdf
CSeq: 56561 MESSAGE
Content-Length: ...
P-Asserted-Identity: sip:sccas.atlanta.com
Content-Type: application/vnd.3gpp.SRVCC-info+xml
```

```
<?xml version="1.0" encoding="UTF-8"?>
<SRVCC-infos>
  <SRVCC-info ATCF-Path-
URI="sip:termsdgfdfwe@atcf.atlanta.com">
    <ATU-STI>sip:sccas.atlanta.com</ATU-STI>
    <C-MSISDN>tel:+1-237-555-1111</C-MSISDN>
  </SRVCC-info>
</SRVCC-infos>
```

The SCC AS updates the STN-SR with the HSS, which in turn transfers it to the MME.

7.2.2. *Session establishment*

The procedure of session establishment is shown in Figure 7.5.

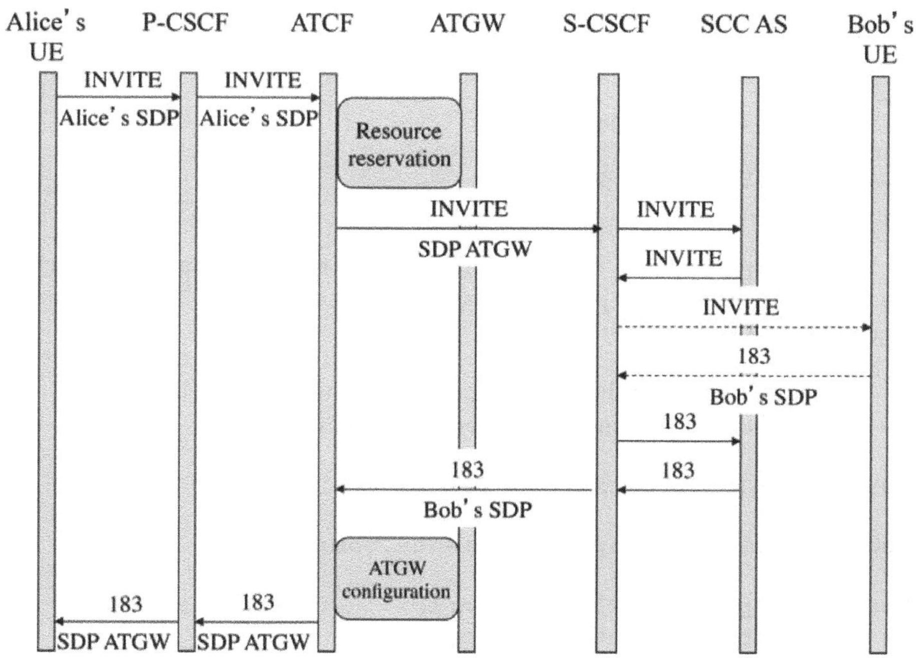

Figure 7.5. *Establishment of the session*

The establishment of the session begins with an INVITE request sent by Alice's UA to Bob's UA.

The headers `Route` contain the identities of the P-CSCF and S-CSCF. It should be noted that Alice's UA does not know the identity of the ATCF, which is, however, situated between the P-CSCF and S-CSCF.

The SDP message body contains the characteristics of the RTP flow of Alice's offer:

– the IPv4 address (`192.0.2.101`)

– the port number (`3456`)

```
INVITE sip:bob@biloxi.com SIP/2.0
...
Route: <sip:pcscf.atlanta.com:7531;lr >
Route: <sip:scscf.atlanta.com;lr>
...
v=0
o=alice 2890844526 2890844526 IN IP4 192.0.2.101
s=-
c=IN IP4 192.0.2.101
t=0 0
m=audio 3456 RTP/AVP 97 96
...
```

As the response `200 OK` to the REGISTER request contained the header `Feature-Caps` with the parameter `+g.3gpp.atcf`, the P-CSCF transfers the request to the ATCF.

The P-CSCF removes the header `Route` containing its own identity and adds the headers `Route` containing the identity of the ATCF.

```
INVITE sip:bob@biloxi.com SIP/2.0
...
Record-Route: <sip:pcscf.atlanta.com;lr>
...
Route: <sip:atcf.atlanta.com;lr>
Route: <sip:scscf.atlanta.com;lr>
...
(sdp)
```

Upon receipt of the INVITE message, the ATCF reserves the resource with the ATGW, which provides the anchor point for the RTP flow, whose characteristics it retrieves.

The ATCF acts as a PROXY SERVER. Thus, it transfers the INVITE request after having removed the header `Route` containing its own identity, and adding that identity to the header `Record-Route`.

The SDP message body transmitted to Bob's UA in the INVITE message is that of the ATGW (on Bob's side), which has replaced that of Alice's UA:

– Alice's IP address (`192.0.2.101`) is replaced with that of the ATGW (`200.0.4.101`);

– the port number (`3456`) is replaced with the value communicated by the ATGW (`8899`).

```
INVITE sip:bob@biloxi.com SIP/2.0
...
Record-Route: <sip:pcscf.atlanta.com;lr>
Record-Route: <sip:atcf. atlanta.com;lr>
...
Route: <sip:scscf.atlanta.com;lr>
...

v=0
o=- 22 333  IN IP4 200.0.4.101
s=-
c=IN IP4 200.0.4.101
t=0 0
m=audio 8899 RTP/AVP 97 96
...
```

The response `183 Session Progress` from Bob's UA contains the characteristics of the RTP flow of Bob's offer, in the SDP message body:

– the IPv4 address (`192.0.2.201`)

– the port number (`6544`)

```
SIP/2.0 183 Session Progress
...
```

```
v=0
o=bob 2890844527 2890844527 IN IP4 192.0.2.201
s=-
c=IN IP4 192.0.2.201
t=0 0
m=audio 6544 RTP/AVP 97 96
...
```

Upon receipt of the provisional response `183 Session Progress`, the ATCF configures the ATGW and replaces the SDP message body received from Bob's UA with that received from the ATGW (on Alice's side):

– Bob's IP address (`192.0.2.201`) is replaced with that of the ATGW (`200.0.4.101`);

– the port number (`6544`) is replaced with the value communicated by the ATGW (`11234`).

```
SIP/2.0 183 Session Progress
...

v=0
o=- 44 555 IN IP4 200.0.4.101
s=-
c=IN IP4 200.0.4.101
t=0 0
m=audio 11234 RTP/AVP 97 96
...
```

7.2.3. *PS-CS handover*

The procedure for PS-CS handover is shown in Figure 7.6.

The decision to effect a handover from the EPS network in PS mode to the UMTS or GSM network in CS mode is made by the eNode B to which the mobile is connected. This decision is made on the basis of the measurement reports transmitted by the mobile in the RRC message *MeasurementReport*.

The eNode B sends the MME the message S1-AP HANDOVER REQUIRED. The handover request may also relate to the flows transferred to the UMTS or GPRS network in PS mode (PS-PS handover).

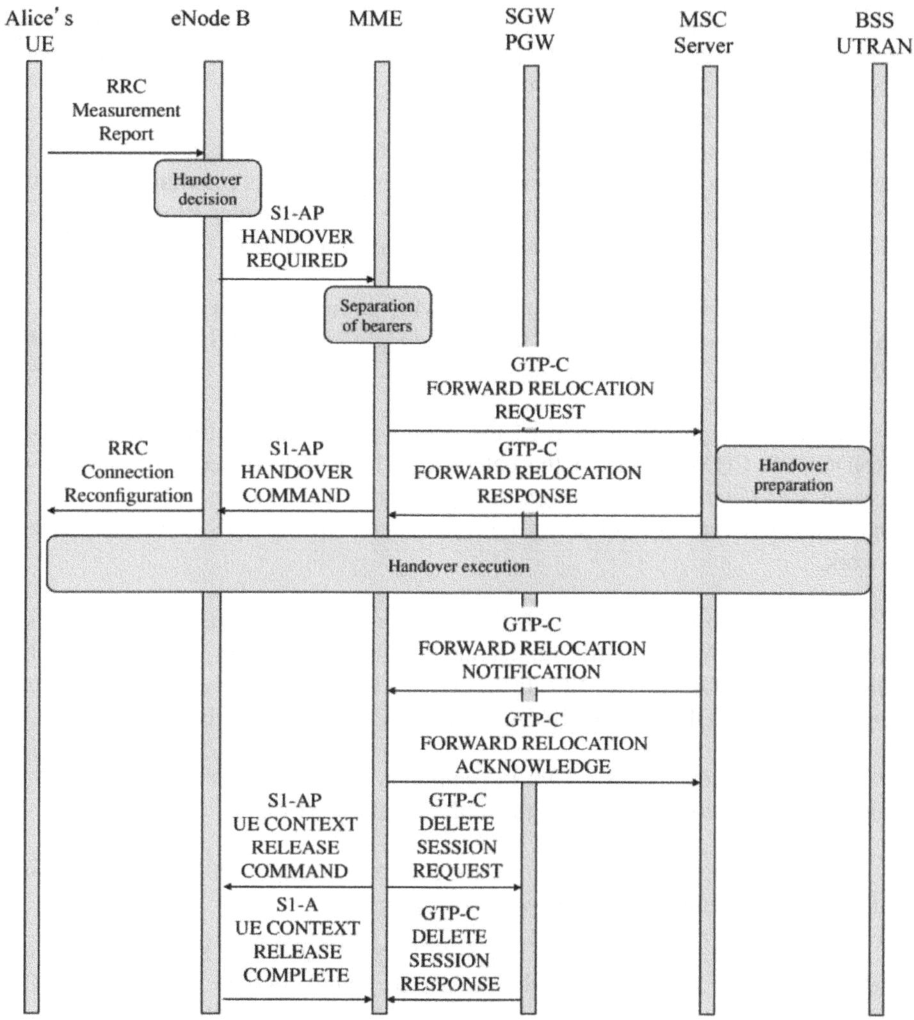

Figure 7.6. *PS-CS handover*

On the basis of the QCI (Quality Class Index), the MME separates the audio flow (QCI = 1) from the other flows and initiates their relocation, either to the MSC Server (RTP flow) or to the SGSN (Service GPRS Support Node).

The MME initiates the procedure of PS-CS handover by transmitting the message GTP-C FORWARD RELOCATION REQUEST to the MSC Server containing the session transfer number STN-SR and the MSISDN (Mobile Station ISDN Number).

The MSC Server transmits a handover request to the UTRAN (Universal Terrestrial Radio Access Network) in the UMTS network or the BSS (Base Station Sub-system) access network in the GSM network.

After having reserved the resources, the access network acknowledges the request received from the MSC Server. The response includes a container which will be sent to the mobile. This container shows the radio characteristics of the 2G- or 3G cell, which will help optimize the mobile connection time.

The MSC Server responds to the MME, indicating that the resources are available for handover, in the message GTP-C FORWARD RELOCATION RESPONSE.

The MME triggers execution of the handover by sending the message S1-AP HANDOVER COMMAND to the eNode B. The eNode B sends the mobile the message *RRCConnectionReconfiguration*, which causes the establishment of the connection between the mobile to the radio station of the BSS or UTRAN access network.

When the mobile is connected, the access network informs the MSC Server, which alerts the MME to the realization of the handover in a message GTP-C FORWARD RELOCATION NOTIFICATION. This message is acknowledged by a message GTP-C FORWARD RELOCATION ACKNOWLEDGE.

The MSC Server assigns a TMSI (Temporary Mobile Subscriber Identity) to the mobile and updates the location of the mobile with the HSS.

The MME triggers the deletion of the contexts in the SGW (Serving Gateway) and eNode B by sending the following messages:

– the message GTP-C DELETE SESSION REQUEST, sent to the SGW. The SGW in turn triggers the deletion of the context at the level of the PGW (PDN Gateway) by sending the message GTP-C DELETE SESSION REQUEST;

– the message S1-AP UE CONTEXT RELEASE COMMAND, sent to the eNode B.

7.2.4. *Transfer of the communication*

The procedure for transfer of the communication is shown in Figure 7.7.

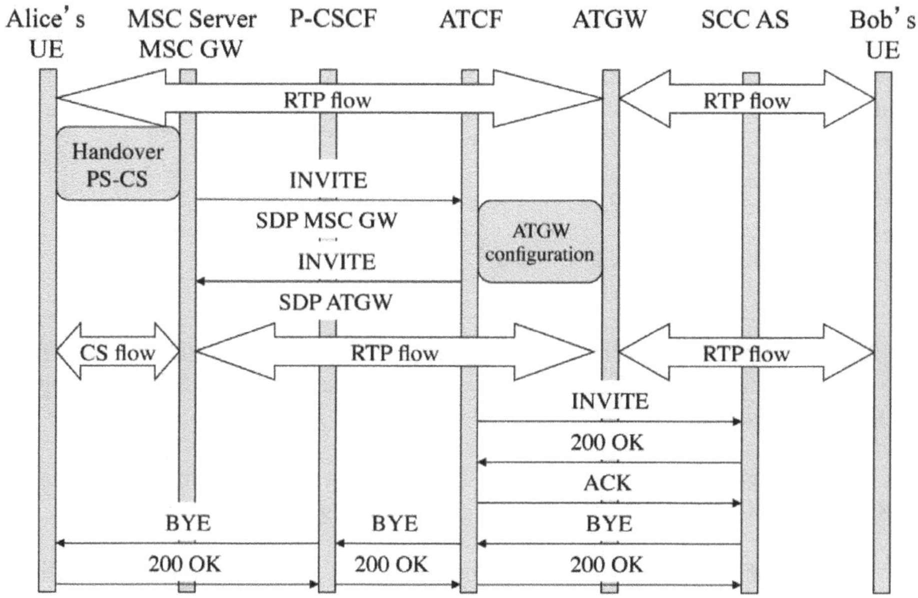

Figure 7.7. *Transfer of the communication*

Alice is in communication with Bob. The ATGW anchors the RTP flow exchanged between Alice and Bob.

Following the PS-CS handover, the CS flow is established between Alice's mobile and the MSC GW. The next phase relates to the establishment of the RTP flow between the MSC GW and ATGW.

The MSC Server sends an INVITE request to the ATCF. The URI of the request contains the STN-SR communicated by the MME during the PS-CS handover.

The header `P-Asserted-Identity` contains Alice's telephone number.

The SDP message body contains the characteristics of the RTP flow. This SDP message was communicated by the MSC GW (IP address and port number).

```
INVITE tel:+1-237-888-9999 SIP/2.0
...
P-Asserted-Identity: <tel:+1-237-555-1111>
...
```

```
Content-Type: application/sdp
Content-Length: (...)

v=0
o=- 2987933615 2987933615 IN IP4 192.4.0.100
s=
c=IN IP4 192.4.0.100
t=0 0
m=audio 5034 RTP/AVP 97 96
...
```

Upon receipt of the INVITE message, the ATCF transfers the SDP message from the MSC GW to the ATGW and responds with the message `200 OK` containing the SDP message sent by the ATGW.

The response `200 OK` incorporates the header `Record-Route` containing the URI of the ATCF and the header `Contact` presenting Bob's identity.

The aim of the next phase is to inform the SCC AS of the transfer of Alice's communication to a CS network.

The ATCF establishes a new dialog with the SCC AS by sending an INVITE request. The URI of the request contains the identity of the SCC AS retrieved at registration.

The INVITE message incorporates the following headers:

– the header `Require` with the value `tdialog` to indicate that the header `Target-Dialog` is required;

– the header `Record-Route` containing the URI of the ATCF;

– the header `Target-Dialog` which specifies that the existing dialog (header `Call-ID`, parameters `tag` in the headers `From` and `To`) must be placed in relation with the dialog established by the ATCF;

– the header `P-Asserted-Identity` containing Alice's telephone number.

The SDP message body contains the characteristics of the RTP flow provided by the ATGW (on Bob's side) when the session was established. As the SDP message body has not changed, there is no need for Bob's UA to perform an update.

```
INVITE sip:sccas.atlanta.com SIP/2.0
...
```

```
Require: tdialog
Record-Route:<sip: atcf.atlanta;lr>
Target-Dialog: me03a0s09a2sdfgjkl491777; remote-
tag=774321; local-tag=64727891
P-Asserted-Identity: <tel:+1-237-555-1111>
...

(sdp)
```

The response `200 OK` incorporates the header `Record-Route` containing the URI of the SCC AS, and Bob's SDP message that the SCC AS had saved.

The SCC AS transmits the BYE request to terminate the initial dialog in the P-CSCF and ATCF, and possibly in Alice's UA if the latter effected a PS-PS handover simultaneously to the PS-CS handover.

If Alice's UA cannot respond with a message `200 OK`, the termination of the dialog takes place on the initiative of the network entities involved.

Bibliography

Recommendations are available on the Website of the 3GPP (3rd Generation Partnership Project): http://www.3gpp.org, and that of the IETF (Internet Engineering Task Force): http://www.ietf.org.

Chapter 1. The EPS Network

[TS 23.401] General Packet Radio Service (GPRS) enhancements for Evolved Universal Terrestrial Radio Access Network (E-UTRAN) access.

[TS 36.300] Evolved Universal Terrestrial Radio Access (E-UTRA) and Evolved Universal Terrestrial Radio Access Network (E-UTRAN); Overall description; Stage 2.

[TS 36.401] Evolved Universal Terrestrial Radio Access Network (E-UTRAN); Architecture description.

[TS 29.274] 3GPP Evolved Packet System (EPS); Evolved General Packet Radio Service (GPRS) Tunnelling Protocol for Control plane (GTPv2-C); Stage 3.

[TS 24.301] Non-Access-Stratum (NAS) protocol for Evolved Packet System (EPS); Stage 3.

[TS 36.331] Evolved Universal Terrestrial Radio Access (E-UTRA); Radio Resource Control (RRC); Protocol specification.

[TS 36.413] Evolved Universal Terrestrial Radio Access Network (E-UTRAN); S1 Application Protocol (S1AP).

[TS 36.423] Evolved Universal Terrestrial Radio Access Network (E-UTRAN); X2 application protocol (X2AP).

Chapter 2. The LTE Interface

[TS 36.323] Evolved Universal Terrestrial Radio Access (E-UTRA); Packet Data Convergence Protocol (PDCP) specification.

[TS 36.322] Evolved Universal Terrestrial Radio Access (E-UTRA); Radio Link Control (RLC) protocol specification.

[TS 36.321] Evolved Universal Terrestrial Radio Access (E-UTRA); Medium Access Control (MAC) protocol specification.

[TS 36.211] Evolved Universal Terrestrial Radio Access (E-UTRA); Physical Channels and Modulation.

[TS 36.212] Evolved Universal Terrestrial Radio Access (E-UTRA); Multiplexing and channel coding.

[TS 36.213] Evolved Universal Terrestrial Radio Access (E-UTRA); Physical layer procedures.

Chapter 3. The CSFB Function

[TS 23.272] Circuit Switched (CS) fallback in Evolved Packet System (EPS); Stage 2.

Chapter 4. SIP and SDP Protocols

[RFC 3261] SIP: Session Initiation Protocol; June 2002.

[RFC 3265] Session Initiation Protocol (SIP) Basic Call Flow Examples; December 2003.

[RFC 3428] Session Initiation Protocol (SIP) Extension for Instant Messaging; December 2002.

[RFC 3262] Reliability of Provisional Responses in the Session Initiation Protocol (SIP); June 2002.

[RFC 3515] The Session Initiation Protocol (SIP) Refer Method; April 2003.

[RFC 6665] SIP-Specific Event Notification; July 2012.

[RFC 3311] The Session Initiation Protocol (SIP) UPDATE Method; September 2002.

[RFC 4566] SDP: Session Description Protocol; July 2006.

Chapter 5. The IMS Network

[TS 23.218] IP Multimedia (IM) session handling; IM call model; Stage 2.

[TS 24.228] Signalling flows for the IP multimedia call control based on Session Initiation Protocol (SIP) and Session Description Protocol (SDP); Stage 3.

[TR 24.930] Signalling flows for the session setup in the IP Multimedia core network Subsystem (IMS) based on Session Initiation Protocol (SIP) and Session Description Protocol (SDP); Stage 3.

[TS 29.209] Policy control over Gq interface.

[TS 29.229] Cx and Dx interfaces based on the Diameter protocol; Protocol details.

[TS 29.230] Diameter applications; 3GPP specific codes and identifiers.

[RFC 3588] Diameter Base Protocol; September 2003.

[RFC 4005] Diameter Network Access Server Application; August 2005.

[RFC 4006] Diameter Credit-Control Application; August 2005.

[RFC 4740] Diameter Session Initiation Protocol (SIP) Application; November 2006.

Chapter 6. Telephone Services

[TS 24.604] Communication Diversion (CDIV) using IP Multimedia (IM) Core Network (CN) subsystem; Protocol specification.

[TS 24.605] Conference (CONF) using IP Multimedia (IM) Core Network (CN) subsystem; Protocol specification.

[TS 24.606] Message Waiting Indication (MWI) using IP Multimedia (IM) Core Network (CN) subsystem; Protocol specification.

[TS 24.607] Originating Identification Presentation (OIP) and Originating Identification Restriction (OIR) using IP Multimedia (IM) Core Network (CN) subsystem; Protocol specification.

[TS 24.608] Terminating Identification Presentation (TIP) and Terminating Identification Restriction (TIR) using IP Multimedia (IM) Core Network (CN) subsystem; Protocol specification.

[TS 24.610] Communication HOLD (HOLD) using IP Multimedia (IM) Core Network (CN) subsystem; Protocol specification.

[TS 24.611] Anonymous Communication Rejection (ACR) and Communication Barring (CB) using IP Multimedia (IM) Core Network (CN) subsystem; Protocol specification.

[TS 24.615] Communication Waiting (CW) using IP Multimedia (IM) Core Network (CN) subsystem; Protocol specification.

[TS 24.616] Malicious Communication Identification (MCID) using IP Multimedia (IM) Core Network (CN) subsystem; Protocol specification.

[TS 24.628] Common Basic Communication procedures using IP Multimedia (IM) Core Network (CN) subsystem; Protocol specification.

[TS 24.629] Explicit Communication Transfer (ECT) using IP Multimedia (IM) Core Network (CN) subsystem; Protocol specification.

[TS 24.642] Completion of Communications to Busy Subscriber (CCBS) and Completion of Communications by No Reply (CCNR) using IP Multimedia (IM) Core Network (CN) subsystem; Protocol specification.

Chapter 7. The SRVCC Function

[TS 23.216] Single Radio Voice Call Continuity (SRVCC); Stage 2.

[TS 23.237] IP Multimedia Subsystem (IMS) Service Continuity; Stage 2.

[TS 24.237] IP Multimedia (IM) Core Network (CN) subsystem IP Multimedia Subsystem (IMS) Service Continuity; Stage 3.

Index

C

D

E

F, G

Printed and bound by CPI Group (UK) Ltd, Croydon, CR0 4YY

13/09/2026

14961076-0003